The Art of Science

•

The Art of Science

Boris Castel and Sergio Sismondo

•

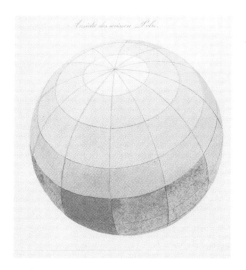

broadview press

©2003 Boris Castel and Sergio Sismondo

All rights reserved. The use of any part of this publication reproduced, transmitted in any form or by any means, electronic, mechanical, photocopying, recording, or otherwise, or stored in a retrieval system, without prior written consent of the publisher — or in the case of photocopying, a licence from CANCOPY (Canadian Copyright Licensing Agency), One Yonge Street, Suite 1900, Toronto, Ontario M5E 1E5 — is an infringement of the copyright law.

National Library of Canada Cataloguing in Publication Data

Castel, B. (Boris)
    The art of science

Includes bibliographical references.
ISBN 1-55111-387-2

1. Science — Philosophy.   2. Science — Methodology.   3. Art and science.
I. Sismondo, S. II. Title.

Q175.C37 2001   501   C2001-930234-7

Broadview Press Ltd. is an independent, international publishing house, incorporated in 1985. Broadview believes in shared ownership, both with its employees and with the general public; since the year 2000 Broadview shares have traded publicly on the Toronto Venture Exchange under the symbol BDP.

We welcome comments and suggestions regarding any aspect of our publications – please feel free to contact us at the addresses below or at broadview@broadviewpress.com.

North America
Post Office Box 1243, Peterborough, Ontario, Canada K9J 7H5
3576 California Road, Orchard Park, NY, USA 14127
Tel: (705) 743-8990; Fax: (705) 743-8353;
e-mail: customerservice@broadviewpress.com

UK, Ireland, and continental Europe
Thomas Lyster Ltd., Units 3 & 4a, Old Boundary Way,
Burscough Rd, Ormskirk, Lancashire L39 2YW
Tel: (1695) 575112; Fax: (1695) 570120
email: books@tlyster.co.uk

Australia and New Zealand
UNIREPS, University of New South Wales
Sydney, NSW, 2052
Tel: 61 2 9664 0999; Fax: 61 2 9664 5420
email: info.press@unsw.edu.au

www.broadviewpress.com

Broadview Press Ltd. gratefully acknowledges the financial support of the Government of Canada through the Book Publishing Industry Development Program for our publishing activities.

Typesetting and assembly: True to Type Inc., Mississauga, Canada.

PRINTED IN CANADA

CONTENTS

· · · · · · · · · ·

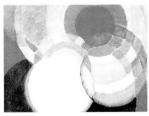

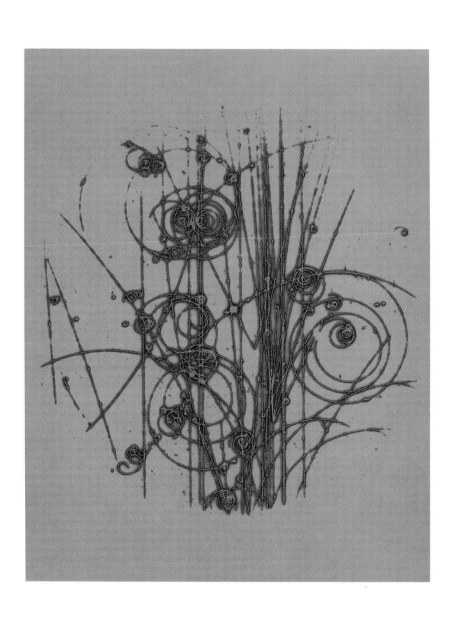

The Computer and the Genius

•

The modern world, and perhaps what it means to be modern, is thoroughly entwined with science. Despite this, most people see science from a great distance and at an oblique angle. We constantly face issues that science brings to our attention, or that science has helped to create. We see science through technological changes, which have created huge leaps in standards of living but have also created pollution, devastating wars, and increasingly difficult medical choices. We see science through the eyes of popular culture, which simultaneously glamorizes and vilifies it. And we see science through memories of high school textbooks, of rote learning, and of confusion in the face of esoteric facts.

This book provides a more direct look at scientists as they work to create knowledge. It presents pictures, stories, and analyses of the "doing" of science, the activity that lies behind articles in newspapers, portrayals on film and in novels, and facts in textbooks. The emphasis here is on activity, not results: no book could hope to explain the current state of scientific knowledge. But a single book can actually do something more interesting and valuable; it can convey some of the flavour of scientific research. This book doesn't try to explain why quarks are the way they are, but why theories about quarks can change, and why

◄ Jacques-Louis David's portrait of the Lavoisiers makes a wonderful statement about the art of science. When this portrait was painted, in 1788, Antoine Laurent Lavoisier had established himself as one of the most eminent scientists of Europe. His experiments, using the chemical apparatuses on his table and at his feet, had justly received wide attention. But they have been relegated to the perimeters of this painting! More prominent is Lavoisier's pen and paper, his tools for artfully representing nature. His articles often were crafted over a period of years, in draft after draft, the central points and arguments being developed in the course of writing.

Lavoisier has even been placed in the shadows, compared to his elegant young wife. Though she often aided him in his experiments, Mme. Lavoisier is here cast in the role of the muse; she represents the "Nature" that her husband is describing in his manuscript. In David's portrait, then, Lavoisier effaces his own contribution even while he asserts it: nature cannot write her own story, but the writer insists on his own secondary role.

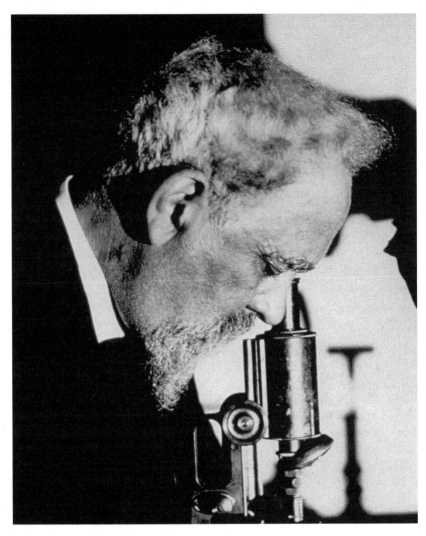

Two observers at work: at right, Pablo Picasso in his 70s, still exploring new forms of expression; at left, physicist Jean Perrin examining the motion of small particles in colloidal suspension.

scientists might not all agree about those theories. Through the development of theories, the work of experimentation, and the craft of argument, we show some of the excitement and some of the drudgery of science, and in the process hope to make the rest of modern scientific work more comprehensible.

At the same time, *The Art of Science* shows some surprising faces of the scientific enterprise. Its primary goal is to show how thoroughly human that enterprise is, to put it within the reach of everybody's imagination. Yet because the humanity of science is so rarely described, even by scientists, some of what it says runs counter to our dominant images. For example, like artists, some scientists have thrived on changing modes of representation: altering the focus, turning background into foreground, and framing knowledge differently. Logic is no more important to science than it is to any other activity. What matters more is how scientists learn to reason about their particular subject matters. Reasoning is a skill, and not a timeless one. Perhaps most surprisingly, experimentation does not straightforwardly reveal nature, but reveals more fundamental structures, which simultaneously underlie nature and yet contain traces of its human origins.

Science is an art. There may seem to be something quite incongruous about juxtaposing "art" and "science." In their original senses, the arts encompassed much more than they do typically today, and the sciences much less. In addition to practices that combined beauty and understanding, everything that now goes under the heading of technology—from the Greek *tekhnë*, art—would have once been considered an art or the product of art: technologies are by definition artificial. The sciences once included only those studies that produced demonstrative knowledge, knowledge that flowed logically from certain premises. Today, the core meanings and connotations of "art" and "science" still reflect their earlier uses. *Arts* depend on the human ability to manipulate and alter nature, so they are human and rely on skill. *Sciences* depend on the human ability to observe and logically analyze nature, so they are nonhuman and depend on logic. *Arts* give new shape and form to familiar materials, making objects that are useful and aesthetically pleasing. *Sciences* do not create new shapes and forms, but study and describe the forms of nature exactly as they occur or are discovered. *Arts* are never final or definitive, always suggesting future developments or changes. *Sciences*, on the other hand, should uncover absolutes or truths. On their surfaces arts and sciences may seem unlike as any activities can be. On their surfaces they are diametrically opposite. Nonetheless, we think that today's common understanding of the connotations of "science" is misleading; that in fact, like the arts, the sciences are human skill-dependent, creative, and rarely final.

Although we certainly do not want to claim that the practice of science and fine art are the same, or even that they are particularly similar human activities, the parallels and differences are intriguing and revealing. The modern physicists who created the quantum revolution made intellectual moves similar to some of their artist contemporaries. Experimenters in the sciences require manual and visual skills, as do painters and sculptors. The social structures of fine art and science are very different—artists tend to work individually, while scientists almost all work for larger institutions—but they have some points of contact in their respective processes of evaluation and validation. And much of our argument is that scientific knowledge is not a mere copy of nature, just as art is not a mere copy of its subjects. For these reasons and more we are happy to talk of "the art of science."

A First Myth of Science: Scientists Are Computers

If we believe popular images, scientists are the most logical of creatures. Like First Officer Spock of *Star Trek* fame—his character replaced by Data in the newer *Star Trek*—they are always measuring, calculating, and classifying. Understanding something, for computer-like scientists, means placing it into a perfect grid of knowledge, finding just enough about it to be able to deduce all else.

Such images come straight from the origins of science. René Descartes (1596-1650) articulated a vision of science as grounded in rigorous logic and some basic mechanistic intuitions. He tried to explain everything physical in terms of matter in motion: collisions, pressure, heat, and inertia. Thus Descartes could write books with titles like *Le Monde* and *L'Homme*, books that systematically demystified and disenchanted the world by mechanizing it through and through. Never mind that we now find much of his work in these books crude, and even false: what is interesting is the surety and logic with which Descartes moved. Starting from the basic physical principles that he believed, he set out to explain the world, to deduce all phenomena. Given his principles, Descartes could believe that nothing in the world would be remarkable in the sense of being unexpected.

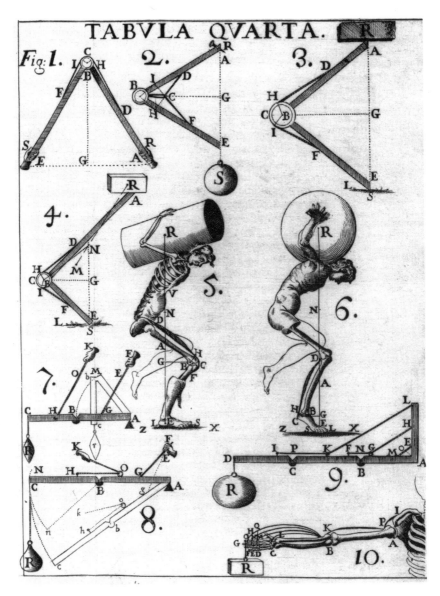

A course in robotics? Giovanni Borelli (1609-1679) strongly believed that the science of mechanics could be applied to human motion, as he demonstrated in the illustrations of his book *De Motu Animalium (On Animal Motion)*. Here we see the influence of a Cartesian mechanical vision of the world.

Another scientist with a computer-like image was the mathematician and astronomer Pierre Simon Laplace (1749-1827). By the age of 24 he was already known as "the Newton of France." That title was not entirely misplaced, for Laplace worked mostly in the areas for which Newton was best known—the calculus, and planetary motions—and he did so within a Newtonian framework. He thought that with Newton's physics and new mathematical tools it would be possible to understand all of astronomy in terms of gravitation and the collision of masses: the Laplacean universe is like a giant game of pool, with each particle a ball, all of them in motion at once, with gravitational attractions between each pair. According to lore, when a reader asked him what place there was for God in such a universe, Laplace answered, "I have no need of that hypothesis."

The idea of scientists as creatures of perfect logic permeates our culture, even in areas that are suspicious of the utility of logic. We can see this idea in the rapid-fire exchange between the android Roy Batty, the "Nexus 6 replicant" in Ridley Scott's dark film *Blade Runner*, and Elden Tyrell, the scientist-entrepreneur who designed him. They confront each other in Tyrell's apartment on the top of the Tyrell Corporation building, a building with sides made to look like massive computer boards.

> *Tyrell:* To make an alteration in the evolvement of an organic life system is fatal. A coding sequence cannot be revised once it's been established.
> *Nexus 6:* What about EMS recombination?
> *Tyrell:* We've already tried it. Ethyl methane sulfanate is an alkynating agent, a potent mutagen, it created a virus so lethal the subject was dead before he left the table.
> *Nexus 6:* Then a repressor protein that blocks the operating cells.
> *Tyrell:* Wouldn't obstruct replication, but it does give rise to an error in replication so that the newly formed DNA strand has a mutation and you've got a virus again.

For every question, Tyrell has an immediate answer, calmly delivered; the strongest emotion in his voice is a little impatience. His admiring employee, Sebastian, smiles at Tyrell's prowess, his ability to match wits with the Nexus 6. In the creation of replicants everything has been thought of, every possibility considered. As a result, Roy, the Nexus 6, is doomed to die young.

In this image the essence of science is rationality. Rationality by itself transforms data into theories, which are then applied to new phenomena. From this vantage point the only real limitations to science stem from the shortage of data and the shortage of time to do all of the necessary calculations. In science, pure rationality does not leave room for real decisions in science, because the correct conclusions are fully determined by the data and within the theories. Real decisions require judgement, because none of the options is completely dictated by the evidence. According to the scientist-as-computer myth, science leaves no room for judgement. And short of finding new data, there is no room for questioning conclusions, because computers only calculate.

The public products of science reinforce the image of scientist-as-computer by presenting the facts of science as always firmly embedded in a huge knowledge network. We see something of the myth of objectivity in the opening line of so many of our second- and third-hand stories of science. "Researchers have discovered..." or "Scientists say that..." begin many newspaper and magazine articles, radio and television reports. Just as often, the researchers are left out of the picture, and the discovery itself is left standing alone, certain and objective. Of course debates and questions often appear further down in the story, but for the moment we are talking only about the myth of the computer, one pole in the imagery of science.

The myth is not a popular one, but is adopted by scientists themselves, at least as an ideal. Exceptional scientific performers display full control over subject matters. Researchers are expected to act dispassionately. Perhaps most notably, students are presented with a picture of established knowledge as a seamless and rigid structure. Textbooks of physics, chemistry, and even biology are organized to leave little room for questions. They display bodies of established knowledge, often interwoven so tightly that each new fact or theorem that the student learns falls neatly into place between, above, and below other facts already learned; the world could be no other way.

The myth of scientist-as-computer is one of pure rationality and perfect objectivity. Although it is clearly a myth, it isn't clear how much of one it is. Over the course of this book we will attempt to delimit the scope of this image, while also showing something of its force. We show how rationality is filled with judgement, and as a result is made more useful. Yet the myth has power as an ideal, something towards which scientists strive. And scientists can seem like computers because they are always creating arguments to convince their colleagues; every bit of research, even research in progress, is infused with those arguments. Thus, while perfect objectivity can only be a myth, scientists often create knowledge without any large and noticeable imperfections. The appearance of perfect logic, then, is the product of much careful work of persuasion.

A Second Myth of Science: Scientists Are Geniuses

No broad and successful institution can have only one public face or one internal ideal. With only one ideal every deviation becomes a failing. Ideals have to exist in some sort of balance, to account for the complexity of worldly actions. Thus, proverbs often come in pairs: "Look before you leap," but "He who hesitates is lost"; "Absence makes the heart grow fonder," but "Out of sight, out of mind." The trick is to know which of the pair to choose for the moment.

The myth of the scientist as computer is balanced by the myth of the scientist as genius. If the first myth represents science as pure objectivity, the second describes pure subjectivity. The first roots the power of science in logical deduction, and the second in insight and intuition—the contrast between Holmes's "Elementary, my dear Watson" and Archimedes's "Eureka!" The second myth perpetuates the romantic image of science, in which genius alone can see to the heart of things to create beautiful and terrifying visions. The genius does not simply know more than other people, but also has the ability to unlock the big secrets of the universe.

The romantic image is crystallized in Albert Einstein's remark that his theory of general relativity was "too beautiful not to be true." Indeed, we might take Einstein as an emblem of the romantic genius, right down to his dishevelled appearance, his

violin-playing, and his amusing antics. Einstein created a theory of space and time with fundamental implications, but it was a theory that few people could understand, and its incomprehensibility certainly has added to its mystique. Einstein could see the possibility of an atomic bomb, wrote a letter to President Roosevelt to urge him to build one, but repented and became a peace activist in horror of what he had contributed to. His humanity was thus both superior and flawed. When Einstein died the pathologist preserved his brain in formaldehyde, creating widespread speculation about that pickled brain, and also a furore—it was an act that seemed to locate Einstein's genius in the form and structure of his brain, suggesting that if we studied that brain thoroughly enough we would know how to create genius.

Let us return to *Blade Runner* for a moment, to see it as a modern *Frankenstein*. Tyrell, the scientist creator, quickly dashes the hopes of the Nexus 6, his creation, but then tries to show some of the beauty of his vision, even if it is a vision without hope:

> *Tyrell:* The light that burns twice as bright burns half as long. And you have burned so very very brightly, Roy. Look at you. You're the prodigal son. You're quite a prize.
>
> *Nexus 6:* I've done questionable things.
>
> *Tyrell:* Also extraordinary things. Revel in your time.

Tyrell shifts mode, from computer to romantic, replacing the dispassionate responses with a more emotional invocation of power and beauty. He adopts the posture of the father, one whose son is the pure product of his mind. The adoption does not save his life, however, for the replicant turns on him and destroys the creator who made him imperfect.

We have called this a "myth" of genius, but not because we want to claim that there is no genius in science, particularly in great achievements. Certainly, just as the computer becomes an ideal for scientists to strive toward, so is genius. We want to show that genius is not perfectly self-contained. There are intellectual moves that allow for innovative and new perspectives, moves that both solve old problems and create new ones. There are *strategies* of genius, ways of doing science—or anything else—that create robust and fruitful new concepts. Few people pursue these strategies, and fewer pursue them well. Those who do may become seen as, and may become, geniuses.

It is no coincidence that the examples so far are of men. Until the past 40 or so years, there were few opportunities for women to join the ranks of scientists, and even today opportunities often are limited in subtle ways; thus men are overrepresented in the history of science, and inevitably in the historical stories told in this book. Furthermore, the myths of the computer and the genius are largely masculine myths, in that they describe qualities stereotypically associated more with masculinity than femininity. Imagine either a cool, clinical scientist, or a mad genius. It is much easier to picture these characters, particularly in their most cartoon-like form, as men rather than women. As our images

Werner Heisenberg (left) and Niels Bohr share lunch and ideas at the Copenhagen Institute around 1930.

become less cartoon-like, and as we forget former stereotypes of men and women, it becomes easier to put a woman's face above the white lab coat.

In popular conceptions of science, we find computers represented as both good or bad, and as neither. The good computer is the servant of humans, providing essential information humans need to live better lives, to better understand the world around them. The bad computer is the technocrat, whose systematic picture of the world veers into totalitarianism. Similarly, the myth of the scientist as romantic genius has both positive and negative incarnations. On the one hand we have beneficent brilliance and insight: genius makes the world a more interesting and beautiful place. On the other hand we have the scientist-genius run to insanity, whose creations are dangerous and terrifying. And according to some stories dangerous creations turn on their creators—narratives of hubris are among the few narratives of science we seem able to create.

• •

Where Are the Boundaries of Science?

When asked to define science or art, most people, including most scientists and artists, find it difficult to answer precisely, but they know roughly what they are trying to define. "Science is about asking good questions," one person says. "In science the key is to work methodically, and never to say more than you know," says another. "Art is all an attempt to represent a transcendent world," and "Artists represent their relations to the things around them." Although such statements disagree with each other, they more or less agree on boundaries. The work of building institutions, selling ideas or paintings, grantsmanship, keeping accounts, maintaining contacts are all peripheral to science or art. But why do we draw the boundaries so narrowly?

Although art may be a vocation, the most successful artists generally are very good at some aspect of self-promotion or the broad business of art. There are even books on how to be a successful artist, covering topics as various as the artist's way of life, preparing a portfolio, taxes, copyright, understanding art collectors, and safeguarding health. These peripherals are helpful for an artist to thrive. If most artists' materials are toxic or hazardous—and scientists face parallel hazards—why shouldn't we see the knowledge and skills to deal with them, such as knowing about the substances, creating proper ventilation, and wearing the right protective clothing also as an essential part of art?

It is part of artistic culture for the artist to appear unconcerned or even incompetent about all of the "peripheral" activities, and the same can be said for scientific culture. Marking a sharp distinction between the core and the periphery in science or art helps mythologize these activities by separating out "pure" realms of activity held to be untainted by politics, poor tools, or other mundane considerations. There is something valuable, something beautiful about that separation. Yet, it obscures much that is fascinating about art and science, and passes over ways even the cores of those activities are connected to worldly matters. The importance of the core realms of science and of art is maintained throughout this book, yet we also try to balance their importance with reminders of their lack of purity. Thus we hope to keep together both what is beautiful and the people who make that beauty.

• •

Beyond the Myths

In this book we endorse neither of the two myths of science, nor do we completely reject them. Both suggest ideals, though nobody can live up to them. They also provide a perspective for viewing individual scientists and particular episodes in science. In Chapter Two, a short study of approaches to innovation in early twentieth-century physics, with some comparisons to innovations in early twentieth-century art, we find elements of the myth of the genius. In Chapter Three we discuss the myth of pure rationality, and show that logic has limited force, and that reasoning skills develop with time and experience. Chapter Four is a look at an ongoing scientific controversy; controversies show the myth of the computer at its weakest, though their resolution erases that weakness. In Chapter Five we look at the practice of science in the laboratory and at the construction of experiments. These activities require discipline and care if they are to produce results; not the discipline of computers, but the discipline that comes from knowing what is needed to convince other people. Thus, the work of science is social from the start. This takes us to Chapter Six, in which we turn to the work of science in a broader social context and look at some institutions that have developed around modern science. In our final chapter we address a question that has become prominent: is an end to science in sight?

Our goal is to move past the superficial understanding of the sciences, embodied in the myths of the computer and the genius, to reach understandings that are thoroughly human, the terms that we associate with other human activities, such as the arts. In the end rationality is made subservient to a more modest reason, objectivity becomes an effect of the hard work of communities,

In stories about scientists a frequent theme is hubris—Icarus flying too close to the sun or Dr. Frankenstein thinking he can create and control life. Both the genius and the computer may fail to understand the limitations of their totalizing visions; if they ignore the subtleties and idiosyncrasies of the world, and of people in particular, they may fail to appreciate the risks inherent in their ideas and intentions. Unbeknownst to the astronauts in Stanley Kubrick's *2001: A Space Odyssey*, the computer HAL is reading their lips as they plot against him. Similarly, the Apollo XIII spacecraft fails, deadly viruses are accidentally let loose, and Jurassic Park stampedes out of control. In stories of hubris, salvation can be found in a genuinely human response to the inhumanity of science. It takes a change of heart or mind—love or heroics—to set nature right again.

and genius looks like a style of intellectual strategy. Most importantly, scientific truths do not have to be final, without being any less truthful; we show that the human dimension of science implies that it can be a never-ending quest. We hope that science will look a little less miraculous, but no less impressive.

CHAPTER TWO

Painting the Invisible

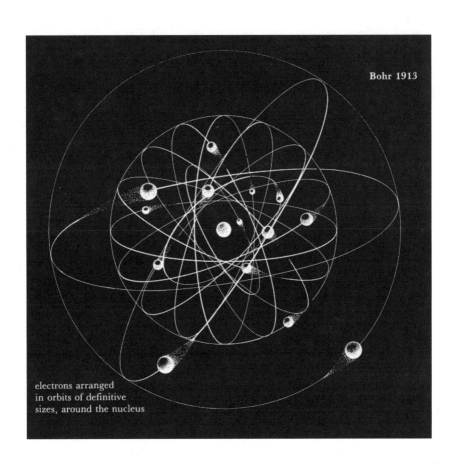

Bohr 1913

electrons arranged
in orbits of definitive
sizes, around the nucleus

●

Using the example of early twentieth-century representations of the atom, we show that science thrives on changes of perspective. Reconceptualizations create new problems and new areas of research. They can have effects that ripple through disciplines and even into the culture at large. Important reconceptualizations allow us to see their creators as brilliant—in a sense, then, successful revolutions create geniuses, not the other way around.

The first 30 years of the twentieth century saw the collapse of what is now called "classical" physics, a term that suggests the closure, or even completion, of a project. In that short time, every major cornerstone of the earlier scientific edifice was overturned. Atoms started the period stable and indivisible, the ultimate constituents of matter. They ended changeable, the decomposible interactions of strange particles. Space and time, initially conceived as independent absolutes, ended fused together and shaped by matter and energy. Energy started out continuous and ended up discrete. Light began as a wave, briefly looked like streams of particles, and ended as something that behaved like both. And more importantly, deterministic causality—the fundamental relation between cause and effect that defined classical physics—appeared to be violated at the most elementary level of nature.

Reflecting on that extraordinary period, Einstein wrote: "All my attempts to adapt the theoretical foundations of physics to this knowledge failed completely. It was as if the ground had been pulled out from under one, with no firm foundation to be seen anywhere upon which one could build." At the outset Einstein had, of course, been one of the leading revolutionaries, but by the late 1920s he was resisting the latest innovations in quantum mechanics. Even Werner Heisenberg, one of the architects of the quantum mechanical revolution, claimed, using exactly the same metaphor, that "the foundations of physics have started moving ... this motion has caused the feeling that the ground would be cut from science." Many physicists must have felt very uneasy.

In this chapter we sketch three well-known developments of the picture of the atom during those 30 years, not because they are typical, but because they show a little of the workings of seismic changes of vision; these re-envisionings of the atom are the hallmarks of genius. But genius does not stand on its own. The dramatic developments of twentieth-century physics also show strategies of genius, the types of intellectual moves that can create rich novelty and new fundamental truths. The seismic revolutions of early twentieth-century physics are the products of those intellectual strategies, strategies that made novel use of the available tools, held fast to anomalies, accepted the paradoxical, and as a result made visible the invisible.

The interleaved vignettes of episodes in the history of modern art hold some surprisingly similar stories. The artists we choose—Cézanne, Duchamp, and Mondrian—were not interested in atoms. But like the physicists we choose—Rutherford, Bohr, and Heisenberg—they were interested in picturing what is not visible. Like their physicist counterparts their achievements involved systematic use and novel recombination of the resources to hand: the achievements of earlier artists, technologies developed for other purposes, and even simple mathematical insights they found useful. These artists approached their tasks in ways that they considered scientific, carefully teasing apart the constituents of images and isolating those constituents upon the canvas. Both artists and physicists, then, adopted radical modes of representation. To demonstrate an essential conception, they had to abandon surface appearances and violate taken-for-granted precepts— precepts of either the painterly art or of physical intuition. So in the early twentieth century physicists and artists were overturning the prevailing common sense and creating new ways of representing the world around them. Both groups drew on and contributed to the spirit of modernity that flourished then.

Some Precursors to the Quantum Revolution

It is never easy to point to firm beginnings and endings of revolutions: there are always earlier prefigurations, however faint, and late resistance, however futile. Natural-looking beginnings and endings depend on later decisions about what is essential to a revolutionary theory or research program and decisions about where the truth lies. We might choose to begin in the years 1895 to 1897, when Wilhelm Röntgen discovered X-rays, Pieter Zeeman showed the existence of a point charge (the electron) in the atom, J.J. Thomson measured the charge of the particles in cathode rays (also electrons), and Marie and Pierre Curie discovered radioactivity. Although by themselves these events did not change conceptions of atoms, they became important to the development of physics: they provided new questions, new phenomena, and new tools for investigation, thereby helping to sustain a tremendous body of experimental and theoretical activity. As one contemporary writing on the "Outlook for Young Men in Physics" said, the developments had "vastly multiplied the opportunities for new discoveries."

If we turn from experimental to theoretical developments we might instead see the work of Max Planck as an essential starting point in the new picture of the atom. Planck was a precise, traditional man, the stereotype of a patriotic, church-going Ger-

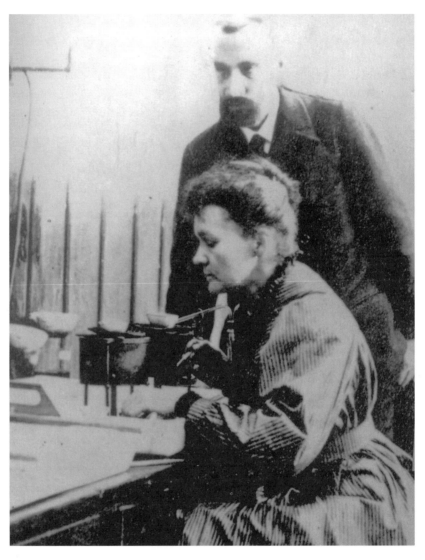

Marie and Pierre Curie in their Paris laboratory in 1906.

man professor. In the year 1900 his study of problems in thermodynamics led him to a formula that could describe the spectra of "blackbody" radiation that experimenters had found. Unfortunately, it was a formula that Planck himself could not quite believe, because it implied that energy was not continuous like a liquid, but came packaged in discrete little bundles, or quanta. However well his equations matched the experimental results, the idea of quanta of energy was too radical for Planck in 1900, and too radical also for other physicists of that time. Planck and others treated the formula and its implications as useful devices, not as literal truth. They became fundamental within a few years.

• •

Cézanne and the Fracture of Space

In 1838, an engineer, Charles Wheatstone, introduced to mesmerized audiences what became known as the stereoscope. The instrument, using mirrors or prisms, enabled the viewer to fuse two slightly different photographs of an object into a single, strikingly three-dimensional image. These stereographic images became common in wealthy European households. The art historian Gerald Needham claims: "No nineteenth-century home of any pretension was complete without a viewer and collection of photographs. The stereoscope was one of the major optical devices of the century, which was enormously widespread, and which played an important part in the changed vision of the later part of the century." The apparatus held by the viewer demonstrated concretely that our Cyclopean world could be deconstructed and analysed as the combination of two underlying "flat" views.

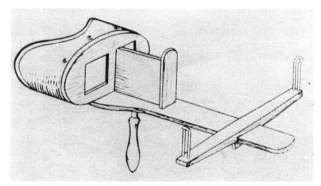

Broken space, combining aspects of two or several points of view on a single canvas is a familiar characteristic of Paul

Cézanne's still lifes. The French painter pioneered that technique in the 1880s and used it so frequently with success that it became one of his long-lasting contributions to modern painting and proved essential to the development of cubism. This success came despite the fact that early on he had few defenders. Regularly submitting his works to Paris competitions, he would choose his most innovative canvases, convinced that their rejection would eventually demonstrate the incompetence of the jury.

Cézanne overrode the tradition of single-point perspective by introducing the idea that a painting can have *multiple points of view*. In his *Still Life with Fruit Basket* (1888-90), his simultaneous adoption of different perspectives is particularly obvious in the discontinuity between the left and right sides of the table, and the views of the vase, jar, and fruit. The art historian Lionello Venturi observes that, in adapting stereoscopic imagery, "Cézanne deliberately distorted objects to represent them from different angles, to turn them around and bring out the fullness of their volume, and to convey by the liberties that he took with perspective, the vital energy of these objects. The beauty of his still lifes, acknowledged by all who see them, depends precisely on the commanding authority with which he convinces us that his 'distorted' vision is truer and more vital than our own."

• •

Rutherford and the Opening of the Atom

Whatever moment one chooses as the beginning of the story of atoms in the twentieth century, Ernest Rutherford's model was certainly the first important complete redrawing of the atom. In 1895 Rutherford left his native New Zealand to take a series of positions at Cambridge, McGill, and Manchester universities. He quickly established himself as one of the most important and skilled experimental physicists of his generation. Rutherford had extraordinary intuition, though not of the sort that physicists call theoretical intuition; he consistently shunned theory in favour of experiment. With the modest means at his disposal, he launched a subtle yet powerful attempt to unlock the mysteries of the elementary constituents of matter—the nucleus of atoms. Modest means were balanced by a healthy ego: when asked late in life why he was always on the crest of the wave of physics, Rutherford replied, "Well, I made the wave, didn't I?"

Pierre and Marie Curie, working in Paris, had shown that the heavy element, radium, spontaneously emitted various kinds of radiation. One of these was known to consist of a stream of electrically charged particles, the "alpha particles," identical to helium atoms when stripped of electrons. These alpha particles did not originate from a helium gas source but were created during the spontaneous decay of the radium atoms. Rutherford saw the possibility of using these alpha particles as probes into the inner structure of atoms. He placed a thin sheet of gold foil in the particles' path to study the scattering pattern that would result when alpha particles hit gold atoms. Instead of the mild deflections to be expected if atoms consisted of diffuse spheres of electrical charge, Rutherford observed that some alpha particles bounced straight back again:

Ernest Rutherford and two graduate students in Rutherford's laboratory at McGill University, 1905.

"It was like firing artillery shells at a piece of tissue paper, and getting some of them returning in the direction of the gun." Rutherford could explain his results only by assuming that most of the atoms' mass was concentrated in a minute, positively charged nucleus, with the light, negatively charged electrons following outer circular orbits, much like the Copernican model of planets orbiting the massive sun. In Rutherford's atomic model, therefore, most of the atom was *empty*. When alpha particles passed through empty space they were only slightly deflected. But when they hit the nucleus they could be scattered in any direction.

Rutherford's achievement stemmed from relatively simple reasoning. But it was a line of reasoning that took both confidence and insight. It took confidence to hold fast to his anomalous experimental result. It took insight to use that result to reject the view of atoms as material bodies, in favour of a view of atoms as systems, Copernican constellations of entities mostly filled with empty space. In short, Rutherford chose a potentially revolutionary approach, rather than one bound to tradition. As it turned out, he was successful.

For 10 years, right through World War I, Rutherford kept on performing his remarkably simple experiments with alpha particles. What he observed was the breakup of nitrogen nuclei occasioned through the collisions with the alpha particles—and the appearance of a new element having all the properties of hydrogen gas. For the first time in history the dream of the alchemist had been realized, and one element had been artificially changed into another. To the officials in charge of the war effort, Rutherford explained his absence: "If, as I have reasons to believe, I had disintegrated the nucleus of the atom, this is of greater significance than the war."

● ●

How Does a Horse Run?

This innocuous question was asked with insistence in 1872 by the horseman and former governor of California, Leland Stanford. To answer it, Stanford hired a British photographic engineer, Edward Muybridge, who had come to prominence for his stereographic photographs of California. Muybridge devised a complex set-up of 24 high-speed cameras to produce a strip of discrete stills, showing distinctions below the level of normal perceptual experience. The results were a complete surprise. Instead of the elegant motion perceived by generations of artists, Muybridge's "gallop" displayed a horse with all four hoofs off the ground at once, a movement that seemed singularly inelegant and "unrealistic."

Meanwhile, the eminent physician, engineer, and natural historian Etienne-Jules Marey had become known in Europe for

Eadward Muybridge, *Daisy Jumping a Hurdle* (1883-87).

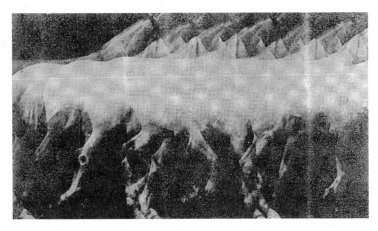

Etienne-Jules Marey, *Walking Horse* (c. 1880).

experiments in animal motion that paralleled Muybridge's studies. By 1882, Marey had devised a photographic gun that rapidly took 12 exposures on one photographic plate. In 1888 he began using moveable film, using first rolls of paper and then celluloid, thus creating a forerunner of today's movie camera. Marey, though, chose to exploit his ideas to expand his studies of space and time.

Through exhibits and articles in popular science magazines, the novel images of Muybridge and Marey were given wide publicity. Even 20 years later they were still fresh enough to prompt painters like Duchamp and the Italian futurists to create some of their most innovative works.

Bohr and the Discrete Atom

Despite his best efforts, Rutherford's Copernican-style atomic models were inadequate to explain the stability of atoms—it would seem they should fly apart. In addition, he could not keep up with the increasing amount of data coming out of his lab. To deal with these problems, Niels Bohr, a young Danish theoretical physicist and a visitor at Rutherford's laboratory, proposed a radical variation on the Copernican idea with a formal simplicity that made it attractive. Electrons would be allowed to occupy certain specific orbits and could remain there without emitting any radiation. Within the atom, then, something about electron space is discontinuous. In Bohr's model, electrons could spontaneously jump from one allowed orbit to another, and in so doing either absorb or emit light. Key to this activity was Planck's idea of discrete packages or quanta of energy, because electrons jump between orbits in a discontinuous way, rather than moving smoothly. Bohr, having matured as a physicist with the idea of quanta, wanted to apply this strange idea to the states of atoms, something most earlier and older physicists could not have done easily.

Bohr's calculation of allowed orbits for the lightest element, hydrogen, with one peripheral electron, fit the known spectrum of hydrogen surprisingly well. Shortly after Bohr's development of his quantum theory, the English experimentalist H.G.J. Moseley developed a formula that unified a large amount of X-ray spectra data, and which matched Bohr's calculations closely, and he announced that he had confirmed the model. With his new-found prestige, the young Bohr began to turn the attention of physicists to Copenhagen, the site of a new theoretical physics institute he was going to direct. Under him, and with almost unlimited support from the Carlsberg Foundation, Copenhagen soon became a European centre of theoretical physics, where increasingly complex atomic theories were elaborated and debated. Bohr's role as a facilitator of physics quickly became more important than his role as a physicist. Mathematically unskilled, he could not keep up with the developments of quantum theory, except in terms of loose analogies. But he had strong intuitions about which ideas and which physicists to support. Given his authority, these intuitions produced a powerful legacy.

Unfortunately, Bohr's simple theory, which had explained so brilliantly the one-electron hydrogen system and had been matched by Moseley's early data, failed completely when confronted with data from heavier elements, where larger numbers of electrons and large configurations of excited states provided a maze of experimental results. Something more was needed.

● ●

Duchamp and the Fracture of Time

The iconoclastic and imaginative Marcel Duchamp developed an intense interest in new styles of painting. Among his sources of inspiration were the photographs of movement by Muybridge and Marey. "In one of Marey's books I saw an illustration of how he indicated fencers and galloping horses with a system of dots delimiting the different movements." That gave him a clue for how to depict a "fourth dimension," which Linda Henderson has shown was key to much art of the period. Duchamp decided that he wanted to produce an exact and scientific foundation for painting motion, to develop a visual language for moving objects, and to transcribe its signs directly onto the canvas. With that in mind he studied geometry, and tried to assimilate discussions of movement and the fourth dimension. The painting he produced in the winter of 1912, the *Nude Descending a Staircase*, was formulated in his new language. It created an immediate uproar, and was considered too revolutionary to be exhibited in the cubist show of that year. Against cubist critics, Duchamp defended his *Nude* as "an expression of time and space seen through the exact representation of time and motion. The reduction of the head in movement to a bare line seems to me defensible. As a form passing through space, it would traverse a line and as the form moved, the line traversed would be replaced by another one—and another—and another.

Therefore I felt justified to reduce a figure in movement to a line rather than to a skeleton; reduce, reduce, reduce was my thought."

● ●

Heisenberg and the Introduction
of Uncertainty

Among the young physicists who most impressed Niels Bohr was Werner Heisenberg, a young German who by the age of 15 had already established a reputation for himself as an imaginative theorist. On Bohr's invitation Heisenberg spent some time in Copenhagen, but despite an invitation to remain there, Heisenberg returned to Germany to study in Göttingen.

In the 1920s the topic of discussion was increasingly focused on understanding atomic structure. What was needed, it was agreed, was a model capable of simultaneously describing the collection of excited atomic states known to exist and the transitions between them, which gave rise to the spectral lines observed. One of Heisenberg's observations on this problem was that quantum jumps needed a non-commutative algebra, that is, an algebra in which A times B is not necessarily equal to B times A. Max Born, an erudite mathematical physicist at Göttingen, recognized that the tools existed, invented some 50 years earlier by a British mathematician, Arthur Cayley. To an older generation of physicists, the mathematics was known as matrix algebra.

In the new matrix scheme, each atom and each physical rule was represented by a matrix, a two-dimensional table of numbers. Atom matrices and rule matrices could then be multiplied by each other to form new atom matrices. For the first time, atomic structure had a genuine mathematical base, albeit one unfamiliar to physicists. In fact, this new "quantum mechanics" allowed people to think of atoms in entirely mathematical terms, leaving the physical behind. Heisenberg's atoms were spare, empty, and purely formal objects, with none of the tangibility of even Rutherford's atoms. Quantum mechanics became very much part of the Göttingen academic culture, steeped in the traditional erudition of German scientific thinking.

Meanwhile, the Austrian physicist Erwin Schrödinger developed an equation that described the same phenomena in terms of waves. Even though it was unclear what was vibrating to create the waves, most physicists initially found the wave approach more

Frantisek Kupka, *Woman Picking Flowers* (1912)

To Frantisek Kupka belongs the honour of introducing first on canvas the idea of multiple images inherited from the new scientific photography. Painted two years before Duchamp's *Nude Descending a Staircase*, Kupka's *Woman Picking Flowers* shows how strongly scientific imagery influenced this innovative painter's style.

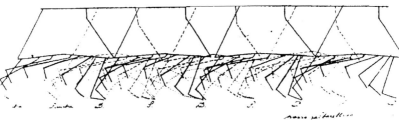

Giacomo Balla, *Young Girl Running on a Balcony* (1912). In the first decades of the twentieth century, the Italian futurist painter Giacomo Balla produced several works inspired by a cinematic type of slow motion. Balla used a stroboscopic effect to impart motion to his model, here his young daughter Luce running on their balcony.

Giacomo Balla, *Dynamism of a Dog on a Leash* (1912)

The same technique was used again to impart a feeling of motion to the bassett hound belonging to his pupil Countess Nerazzini.

familiar, more comprehensible, and more tractable mathematically. This spurred the Göttingen group, and Heisenberg in particular, into a competitive frenzy. Paper after paper was published, in an attempt to find something that the matrix scheme could do that the wave approach couldn't. Although the competition was in vain—Schrödinger's and Heisenberg's versions of quantum theory turned out to be equivalent—it resulted in innovative thinking and much new physics.

Most famously, Heisenberg discovered one of the most dramatic results of the new quantum mechanical world. The ambitious Heisenberg wanted to undercut not just Schrödinger's equation, but the whole picture on which it rested. Thus he engaged in discussions on the foundations of physics with many of the leading researchers in Europe, including Bohr, Wolfgang Pauli, P.A.M. Dirac, Pascual Jordan, and others, drawing on every resource he could find to oppose the wave picture of matter. Pauli, thinking about his large contributions to Heisenberg's ideas, later came to lament his own "conservatism." The result was the famous Uncertainty Principle, which stated that the exact position and the precise velocity of elementary particles could not be determined simultaneously. The importance of this to the development of quantum mechanics was enormous, affecting physicists' interpretations of all of their theories. It also opened up holes in ideas of causality. Here is why: because velocity and position cannot be simultaneously specified, if two electrons could be sent in the same direction with identical velocities, they would not necessarily end up in the same place. So, in the language of classical physics, the same cause could produce different effects. The classical principle of deterministic causality was violated at the most elementary level of matter.

Interestingly, according to the historian Paul Forman, the way for questions about causality had been paved by outside events. Classical determinism had become an issue in the German popular imagination because of the 1918 publication of a monumental treatise, *The Decline of the West*. The author, Oswald Spengler, an impoverished high school teacher, argued that science's mechanistic and reductionist attitudes were major factors responsible for the ultimate decline of Western civilization. In the devastation that followed World War I in Germany, Spengler's pessimism found a ready audience, and the public reception of his millennial prophecy was phenomenal: within five years, the first edition went through 30 printings. By the mid-1920s the climate in Germany had turned anti-scientific, and this affected the scientists themselves. Deprived by the change in public values of the prestige they had enjoyed before and during the war, and deprived by economic collapse of much of their funding, German physicists may have felt pushed to alter their thinking in order to recover a more positive public image. In particular, Forman says that "many resolved that one way or the other they must rid themselves of the albatross of causality," something that quantum mechanics eventually would promise. Forman claims that "the program of dispensing with causality was, on the one hand, advanced quite suddenly after 1918 (and the publication of the *Decline of the West*) and, on the other hand, that it achieved a very substantial following among German physicists *even before* it became 'justified' by the advent of Heisenberg's acausal quantum mechanics." That is, many physicists were willing to abandon determinism before there were good reasons to do so. Substantive problems in atomic physics may have played only a partial role in the genesis of acausal interpretations of quantum mechanics.

• •

Mondrian and the Geometry of Nature

One of the most original tendencies of modern art has been the development of novel approaches to geometrical forms. Probably the most original and successful painter of the twentieth century to develop a personal geometric vocabulary was the Dutch artist Piet Mondrian. Attracted to Paris in 1910 when cubism was in its infancy, the young Mondrian became influenced by Picasso, Braque, and Léger. But he wrote, "Gradually I became aware that Cubism did not accept the logical consequences of its own discoveries. It was not developing abstraction toward its ultimate goal, the expression of *pure reality*.... To create pure reality plastically, it is necessary to *reduce* natural forms to the constant elements of form and natural color to

primary color ... plastic art discloses what science has discovered; that time and subjective vision veil the true reality." At the outbreak of World War I Mondrian returned to Holland, where he remained for the duration of the hostilities. He settled in a secluded seaside town whose flat horizon and small vertical pier formed a grid, which was to become the hallmark of his work.

Mondrian was obsessed with the idea that volumes and even surfaces retained a representational character and had to be destroyed to reach *pure abstraction*. "I came to the destruction of *volumes* by the use of the *plane*. This I accomplished by means of lines cutting the planes, but still the plane remained

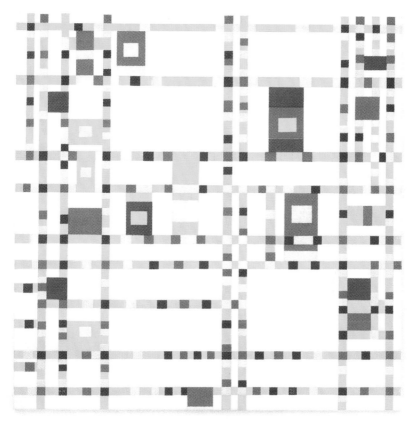

Piet Mondrian, *Broadway Boogie-Woogie* (1942-43) above.
Piet Mondrian, *Composition C* (1934) opposite.

too intact, so I came to making only lines and brought the colour *within* the lines." Now the only problem was to destroy these lines also through mutual oppositions. The "true reality" that Mondrian sought to discover was based on a search for balance and equilibrium resulting from the employment of exact opposites in painting; vertical versus horizontal line, the use of the primary colours (red, yellow, and blue) in opposition to the non-colours (black, white, and grey). These were the elements from which Mondrian fashioned his mid-life paintings, such as his *Composition C*, which many critics consider to be his best.

Casimir Malevich, *The Fourth Dimension* (1924).

The idea of space and higher dimensions influenced a whole generation of painters. Among them Casimir Malevich has probably been the most articulate (and the most colourful) in his inspiration.

Some Lessons: Radical Reconceptions and Strategies of Genius

There is something misleading about these stories of modern physics and modern art. We have provided simple narratives, illustrating revolutions through the achievements of a few researchers and artists. We have not tried to reveal the false paths down which yet other people were travelling, or display the many connections between the efforts of that small number and the efforts of others working at the same time on the same or related issues. However, it should be said that early twentieth-century physics in Europe was an intensely social activity, in which researchers kept up with each other's work through conferences, collaborations, lecture series, and extensive correspondences. They formed close communities and depended heavily on each other. The achievements of Planck, Rutherford, Bohr, and Heisenberg—and of Cézanne, Duchamp, and Mondrian—are impressive in any context, but when we strip them of context we reinforce the myth of genius, by making those achievements appear to stand entirely alone. Towers in a field impress one quite differently from towers in a city.

Context aside, the re-descriptions of the atom and its energy, all of them revolutionary, show what we are calling "strategies of genius" at work. Strategies of genius transform ability into brilliance, hard work into creativity, the familiar into the novel. They become heuristics for finding new types and levels of description, levels that made derivative what had been fundamental and vice versa. Hence, scientific discoveries have profoundly altered our world views. We no longer assume that matter is made of *recognizable* and indivisible entities as was the case at the dawn of the twentieth century. Parallels appear in the work of modern artists, many of whom abandoned all attempts at representing reality in terms of recognizable objects.

Scientific revolutions coalesce around views that overturn established orders. To start these revolutions requires transforming anomalies into foundations. Max Planck created an anomaly in a formula, which could perfectly account for blackbody radiation: the formula had the odd consequence that energy was discrete, not continuous. Planck, the solid, exceptionally able theorist, was for years convinced that while his formula might hold it could not be completely true, because it led to an unacceptable consequence. Niels Bohr, however, accepted the anomalous consequence as foundational—as had Albert Einstein for his own purposes—and designed an atomic model around it. The result was the first quantum atom. We can praise the intuition that led Bohr to accept Planck's work, but at some level he was accepting what had already been well established. Where he differed from Planck was in choosing to see discreteness not as unpalatable but possibly as a characteristic feature of the very small.

Something similar happened when Cézanne picked up the multiple points of view of the stereoscope, which was, after all, a device designed to maintain a single natural point of view. Cézanne saw that multiple points of view were not necessarily unacceptable in art; in adopting them he simply abandoned the tradition of the stationary viewer. When Duchamp absorbed Muybridge's and Marey's photographic discoveries, he was able to depict motion in an entirely novel way, likewise violating established traditions of representation.

Rutherford had enough faith in his own abilities as an experimenter—he did not lack self-confidence—that when he saw the anomalous scattering of alpha particles, he knew that something other than his results had to give way. If the classical model of the atom, pictured as a homogeneous mass, could not account for the occasional direct reflection of alpha particles, then the classical model had to be replaced by one with concentrated masses and empty space. In the process, Rutherford employed another proven scientific strategy: taking something thought to be homogeneous and reconceptualizing it as having parts and structure. Since Rutherford, and up to the current day, exploration of the structure of atoms has revealed more and more complexity. There is perhaps something of a dialectic of simplicity and complexity. Movements often start with dramatic simplifications, but those simplifications are only adequate for a short time and limited range of circumstances. One might even see this in the work of individual scientists and artists: Piet Mondrian's early work reduces the visual complexity of naturally perceived images into straight lines, blacks, whites, and primary colours. His later work keeps those elements, but they are com-

A young Werner Heisenberg

bined much more playfully to capture a broader range of emotions and topics.

Werner Heisenberg helped to unify quantum theory by turning atoms into mathematical objects. He took an old tool and put it to new purposes, yet another useful and recognizable strategy. Heisenberg's representations of atoms succeeded even while leaving behind the physical and spatial intuitions that had been built up around them. The physicality of earlier images gave way to austere matrices, which allowed measurements of the atom to interact in a mathematically homogenous way. More clearly than his predecessors, Heisenberg created a new form for representing atoms; the resulting pictures wrenched them out of the physical world and placed them in a mathematical one.

Of course Heisenberg was also a leader in reconceptualizing not only atoms but also physicists' relation to them. The Uncertainty Principle implies a strategy of stepping back to recognize that the measurement of the minute is not neutral and that experimental data do not transparently represent a world beyond. By so doing, Heisenberg recognized that at the level of atomic physics classical causality might be an unwelcome imposition on the part of researchers, not something inherent in the nature of things. German physicists may have been especially well prepared by intellectual fashion and economic circumstances to challenge classical causality.

In their commitment to representing unseen realities, modern artists and scientists have followed similar paths. Faithful representations do not necessarily make features of the natural world more comprehensible, and they are in no sense mere copies of their subject matters. Rather, a good model, theory, or painting should contain some structural features that make what it describes more interesting, elegant, and/or universal. If theories or paintings can do some or all of these things successfully, then they eventually become regarded as realistic. What once may have been blasphemous representationally becomes faithful to another perception of things. A good representation is true, but truth can only be expressed in human terms.

CHAPTER THREE

Logic and the Construction of Reason

Scientists have to learn how to think about particular subject matters. Scientific reasoning on a topic, then, is not invented once and then static forevermore. Rather, it develops over time, becoming attuned to idiosyncrasies of particular subjects through careful work, and sometimes through fierce controversies. In the end scientific reasoning has the solid and rigid feel of logic, but not the purity of logic: it is, as a result, much more useful.

Scientific Reasoning is an Art

If scientists were computers then their thinking would be flawless and straightforward, just a matter of grinding through the implications of data to arrive at logically necessary conclusions. Of course, scientists are only human, even if popular images, and scientists' own imitations of those images, make them seem somewhat dispassionate, objective, and mechanistic. Being human might sound like a limitation, but in fact it is not. Science cannot be done by mere computers: developing scientific knowledge requires skills that computers cannot have. Reasoning is an art, and reasoning about the natural world is the art that lies at the base of science.

As people learn how to think clearly about particular topics their patterns of reasoning become more and more rigid, and look more and more like computations; this is the truth in the myth. That rigidity, however, is a never fully realized end point in the process of exploration, not a starting point. It is a result of hard-won consensuses about the best data, the best tools, and the

◄ *Reason Reveals the Truth.* This engraving was the frontispiece for the *Encyclopedie*, the central work of the French Enlightenment. Reason here wears a regal crown. Beneath her cascade her allies: Theology (carrying a Bible), Philosophy (with a flame of the spirit), Astronomy (with a wreath of stars), and each of the sciences personified.

A pendulum oscillates in a constant plane. Leon Foucault (1819-68) used this principle to demonstrate the rotation of the earth. Here, in the Pantheon, Reason watches the simple demonstration from above, as if to claim credit.

best interpretations. Once the scientific community has reached substantial agreement on some issue, it has become extremely difficult to locate gaps in its thinking, extremely difficult to question its patterns of reasoning. And without easy openings for questions, that socially produced reasoning takes on the force of computation. In the end, then, is the myth of the computer plausible because science is a social activity?

Scientists Learn to See the Obvious

The introduction to *Princeton Problems in Physics, with Solutions* begins:

> No one expects a guitarist to learn to play by going to concerts in Central Park or by spending hours reading transcripts of Jimi Hendrix solos. Guitarists practice. Guitarists play the guitar until their fingers are calloused. Similarly, physicists solve problems. And hopefully, physicists practice solving problems until doing so seems easy. (Then they find harder problems.)

Why the emphasis on problems? Why do typical textbooks have problems at the end of each chapter, or questions that the readers are supposed to answer, or at the least provide a variety of examples? By using a theory or a fact in a variety of different contexts, students can come to understand how it applies to those contexts, and can transfer their problem solving skills to other contexts. In using a theory students come to understand the range of applicability for what they have learned. Most often we learn the meaning of words by hearing them, reading them, using and abusing them and being understood or corrected, until we know their ranges. In the same way we learn the meanings and ranges of ideas. Thus, the student of physics understands that the watchspring and the rocking horse both are harmonic oscillators and learns to see harmonic motion in a variety of objects. Skill in recognition must be learned, because a definition alone cannot allow one to recognize harmonic oscillators in all of their variety. Once a person has learned to perceive in terms of theoretical concepts, he or she has learned much of what it means to reason with those concepts.

A Brilliant Idea with a Difficult History? The Case of Natural Selection

In 1859, Charles Darwin published his *On the Origin of Species*, containing a brilliant idea to account for the diversity and complexity of living organisms, and their adaptation to their environments.

An elderly Charles Darwin (1881) stands secure in his achievements. Darwin's determined development of his early ideas resulted in a profound theory, and his mobilization of a network of scientific colleagues put that theory in the fore-front of biological research.

That idea, which one writer has rather audaciously called "the single best idea that anybody has ever had," was natural selection. Here is the core of the concept of natural selection, roughly in Darwin's terms: Many more organisms are born than survive to have offspring. Those that manage to reproduce will pass their characteristics on to future generations, since offspring generally resemble their parents. Therefore, characteristics that help organisms survive and reproduce will tend to become more common, at the expense of those that do not. The most beneficial variations, then, will be selected for, in the sense that those are the ones most likely to be found in the next generation. Over thousands and millions of generations one could expect to see large changes, wholly new types of organisms with well-developed features that once were barely present. Initially random variations, subjected to constant pressure, produce order.

Altogether, natural selection is a simple and profound idea. Nonetheless, selection has had a rocky reception over the past 140 years. After some initial enthusiasm, in the decades around 1900 biologists looked to alternative theories to explain organisms' adaptations to their environments and to explain features that biologists didn't see as adaptations. In 1900 Darwinism looked as though it was on its way out, either a failed or very limited theory, though there was no good alternative to replace it. If we imagine science itself in Darwinian terms, natural selection was a theory that was being selected against. Darwin and others had established the evolution—the change through time—of organisms, but many biologists developed other explanations to account for evolution.

By the 1940s the tide had turned again, and it would not be long before essentially all evolutionary biologists were Darwinians. Today, many biologists would say not only that natural selection is true, but that it is necessarily true, and is essentially the only principle needed to explain the entire history of life on earth. So here is a puzzle: Why, if natural selection is now so obvious and so central, did all biologists not quickly become Darwinians?

There are many reasons. Some people had aversions to the materialistic framework of natural selection, the fact that it explains life without any reference to spiritual or immaterial forces. Some people had aversions to the overly vicious picture that they thought Darwinism created, a picture of nature as a "war of all against all." These aversions tended to be associated with religious feelings against Darwinism or hopes for morals from nature. For the most part, though, biologists did not have religious or moral reasons for doubting Darwinism. Rather, they thought there were cases that the theory of natural selection couldn't handle, problems that it couldn't solve, and other theories that could address those cases and problems. For natural selection to become accepted, and the ultimate explanation for evolution, biologists needed to agree about how to solve problems. This meant agreeing about the terms in which to see the biological world.

• •

Lord Kelvin's Objection

When Darwin was trying to put evolutionary biology on a solid scientific footing, many other disciplines were much further developed. In particular, though thermodynamics was relatively young, it had some very solid results and theories. Those results and theories posed an easy challenge to Darwinism. According to thermodynamic arguments put forward by the engineer Fleeming Jenkin and the physicist Sir William Thomson (later Lord Kelvin) in 1867, and developed further by Thomson shortly afterward, the earth was probably only a few million years old. As a hot body in cold and empty space, it would cool down quite quickly, radiating heat outward like a hot rock in ice water. Although it wasn't possible to estimate precisely how old the earth was, according to Kelvin's calculations it was 100 million years at the most, and during the largest portion of that time it would have been far too hot to sustain life.

Darwin needed far more than 100 million years for his slow and steady processes to work. There was good geological evidence for a longer period, and there was plenty of good biological evidence that even if natural selection wasn't correct, some other gradualist theory of evolution was. But both geology and biology were weaker sciences than physics in an important sense: they had much lower levels of status and public esteem. So until evolutionists and geologists could find a mistake in Kelvin's argument, they couldn't dislodge his objection—this was one of the reasons some biologists searched for faster mechanisms of evolution.

In the first decades of the twentieth century, the discovery of radioactivity changed the calculations of the thermodynamics of the earth. With radioactivity, and not just heat, as a source of energy, it would take the earth much longer to cool than Kelvin had supposed. So under the influence of new developments, arguments in the lower-status sciences became stronger than those of their higher-status cousins.

• •

Conflicts of Vision: Does Evolution Have Trajectories?

Particularly telling is the fact that even Darwin did not think that his theory was all-sufficient. Although he thought that natural selection was the dominant mechanism governing evolution, he allowed that some other mechanisms were also important. This is because he saw substantial problems facing his favourite idea.

Natural selection explains how *useful* variations are selected and passed on to future generations, and how *harmful* variations are selected against. But what about variations that are neither useful nor harmful? In 1865 Carl von Nägeli argued that there were many characteristics that distinguished particular species but were of no use to their bearers. That is, processes of evolution had created new species not on the basis of adaptations, but on the basis of irrelevant characteristics. In *The Descent of Man*, Darwin accepted that this was a serious problem. There he said:

> I now admit, after reading the essay by Nägeli ... that in the earlier editions of my "Origin of Species" I perhaps attributed too much to the action of natural selection or survival of the fittest. I have altered the fifth edition of the "Origin" so as to confine my remarks to adaptive changes of structure.

Darwin went on to predict that biologists would eventually discover adaptive value in structures that they then saw as useless,

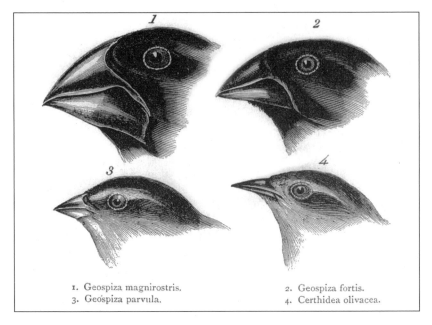

1. Geospiza magnirostris.
3. Geospiza parvula.

2. Geospiza fortis.
4. Certhidea olivacea.

Darwin's finches have become an emblem for evolution. The different species of finches on the isolated Galapagos Islands were adapted to a variety of niches that elsewhere were filled by other birds. Darwin argued that the Galapagos finches had strayed from the South American continent and had slowly diversified on the islands.

but nonetheless recognized in non-adaptive structures one of his "greatest oversights."

The competing theories of orthogenesis addressed the issue of non-adaptive characters, and also explained some apparently non-Darwinian patterns in the fossil record. The great American paleontologist Henry Fairfield Osborne, for example, claimed that forces inherent in particular species propel their evolution along certain pathways or trajectories. On this theory evolution is likened to individual development; the history of a lineage is like the opening of a flower or the maturing of an animal.

Why would anybody believe in orthogenesis? The ancestors of horses are well documented in the fossil record and appear to grow larger, and more like today's horses in other ways, with each evolutionary change. How could selective pressures—pressures like predation, nutrition, and climate—consistently push

in the same direction over many millions of years? Many paleontologists and other biologists saw this, rightly, as a difficulty for
the theory of natural selection and developed an alternative: if
there were among other things an enlarging force inherent in
horses, directing their variation, we would see smaller species of
horses constantly replaced by larger ones. Instead of random
variations, which would result in both larger and smaller horses,
the supposed orthogenetic force would create a slightly higher
proportion of larger horses in each generation. According to
some versions of orthogenesis, eventually horses will become so
large that they will become extinct, being too poorly adapted to
survive.

There is something noble about the theory of orthogenesis,
because it claims that through evolution species approach their
ideal forms—for this reason Osborne called his particular version of the theory "aristogenesis." The theory claims that
species will evolve in more or less predetermined manners,
according to their different natures. The theory of natural
selection, on the other hand, imagines species as plastic, slowly
changing to fit their environments through the accumulation
of chance improvements. If a lineage changes in a constant
direction, it must be steadily becoming better adapted to its
environment, or tracking a steadily changing environment. In
general, both of these would be unexpected over long periods,
though expected over short periods. There are plausible exceptions, but they are probably not common: cheetahs and gazelles
have each become faster because each is a key part of the
other's environment.

Clearly, Darwinians won the debate eventually, though not
because orthogenesis has no plausibility. Modern variations on
the orthogenetic theme continue to be developed, though now
as questions about the right interpretation of Darwinism, rather
than as competitors to Darwinism. Stephen Jay Gould, the well-
known writer and paleontologist, has argued that there are long-
term patterns in the fossil record, and that there are structural
constraints to adaptation by natural selection. Both of these
could be orthogenetic claims if they were central to a theory of
evolution, but Gould is a thorough Darwinian. How did natural
selection triumph so pervasively?

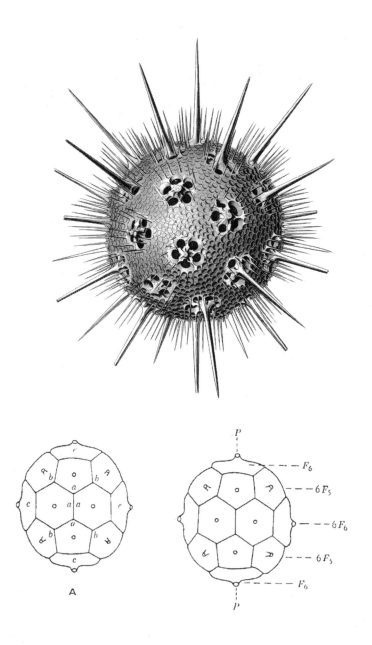

The nineteenth-century biologist Ernst Haeckel was so taken by the perfection of these spiky radiolarians that in 1860 he wrote: "I fell on my knees in front of the microscope and cried out in joy as I offered my most heartfelt thanks to the blue sea." Forty years later D'Arcy Wentworth Thompson included radiolarians in his study and provided a more mathematical description of their structure.

• •

How much can chance do?

Some of the difficult cases for Darwin turned on questions about the power of natural selection. Can chance variations and imperfect selection create the spectacular things that we find in the living world? The nineteenth-century physicist and philosopher John Herschel called natural selection the law of "higgledy piggledy." How could natural selection create an eye, let alone the eye of a hawk? Why would humans evolve abilities for abstract mathematics or poetry, neither of which would aid them in survival under primitive conditions? With all of its dependence on randomness, natural selection sometimes looks incapable of explaining the most exquisite features of the natural world, and so biologists developed other theories to explain these features.

Not only physicists reacted against Darwin's emphasis on chance and history. The biologist D'Arcy Wentworth Thompson's *On Growth and Form* is an intriguing and very widely read anti-Darwinian treatise. Although it was published in 1917, *On Growth and Form* has been in print ever since and continues to provide some hope to opponents of Darwinism.

Thompson disliked Darwin's style of historical explanation, which emphasized the accumulation of small differences. In its place he wanted to put a physico-mathematical style of explanation that identified mathematical patterns in the development and shape of organisms and their parts. The gorgeous shells of many marine organisms, the pattern of seeds on a sunflower, and the pattern of development of mammalian skeletons all are describable in mathematical terms. The illustrations that fill *On Growth and Form* make this point visually over and over again.

Do we need to explain these patterns through natural selection? Thompson firmly believed that we do not. According to him, these mathematical patterns, linked to physical principles of growth and evolution, are themselves explanations. Physics had long assumed that, as Galileo said, the book of nature was written in the language of mathematics, so for Thompson his mathematical structures provided a better account than Darwin's historical ones.

• •

Consensus and Exclusion: The Evolutionary Constriction

Natural selection displaced orthogenesis not with one knock-down argument, but rather by many different lines of argument, all of which contributed to the downfall of orthogenesis. Here are a few.

First, as Darwin predicted, biologists found adaptive value in features they formerly thought valueless. Close study shows that even apparently trivial differences in patterns on snail shells serve as specialized camouflage in different environments, that subtly different plant varieties thrive in subtly different environments, and so on. Darwinians have been so successful that we should probably assume that any given biological feature is an adaptation until we learn otherwise. The burden of proof is now on the non-adaptationist.

Second, orthogenesists did not discover any mechanism that would direct evolution, or even direct variation. Variation was a poorly understood topic until the 1940s, when well-developed versions of Mendelism finally became wedded to Darwinism. Until then neither theory had a good handle on novelty in evolution and whether novelty appears in any regular way.

Third, some of the large-scale patterns that paleontologists thought they had seen turned out to be more the result of the imposition of order by researchers than the result of order inherent in the fossils. In 1944 the paleontologist George Gaylord Simpson argued that oversimplified arrangements of too few fossils created the apparently straight-line development that orthogenesists had seen. Horses did evolve towards their modern shapes and sizes, but not in nearly as straight a line as once was thought. Where patterns can be seen, not all are genuine. Darwinists learned to see disorder in the fossil record, where order had been assumed.

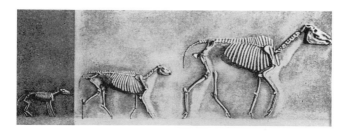

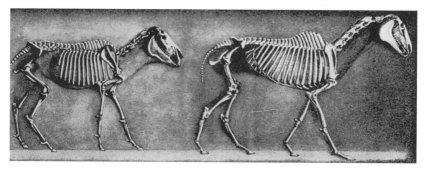

From Frederic Brewster Loomis, *The Evolution of the Horse*, 1926.

Finally, biologists acquired an appreciation of the immense stretches of time that natural selection has had to work. They developed judgements about the rates at which natural selection can have effects, drawing on experience of variability and experience of the pressures of life in the wild. Such judgements about rates are difficult because evolution does not happen on a human time scale. A further step in the solution of the problem is to thoroughly understand the incremental nature of selection. Natural selection creates marvellous features because each step towards such features confers some advantage. By stages over thousands and millions of years, an organ or characteristic can go from being only marginally useful—but still useful!—to being exquisitely fine-tuned. Many biologists found it a hurdle to recognize that an imperfect eye would still be useful to its bearer and could be the basis for further developments.

Logic was not at issue in the dispute we have just described. Darwinists and orthogenesists agreed on the principles of logic, and most of them even agreed that natural selection was responsible for some, but not all, of evolution. They disagreed about whether par-

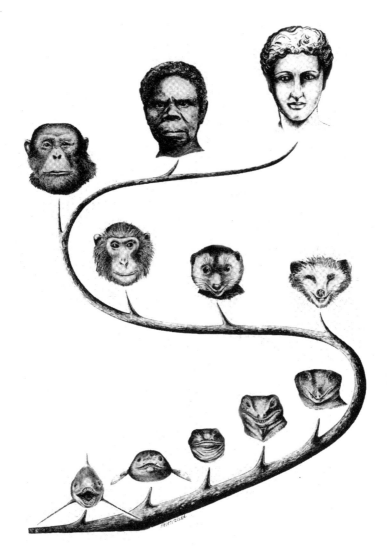

From William K. Gregory, *Our Face from Fish to Man*, 1925. Although this is not explicitly intended to illustrate a teleological view of facial features—towards an ancient Greek ideal of male beauty—it suggests a vision of progress that permeates Gregory's book.

ticular characteristics were adaptive or non-adaptive. They disagreed about whether there were patterns in the fossil record. And most importantly, they disagreed about whether the apparent difficulties faced by each theory were important or trivial. In general, logic could not solve these problems, though experience in reasoning about organisms could, and fairly decisively on the side of the Darwinists.

Thus the theory of natural selection requires not only an insight but habits of thinking, or patterns of reasoning. The standard image of Darwin as having a single grand insight tells only a small part of the story; even Darwin accepted alternative mechanisms to natural selection, alternative mechanisms that biologists today do not generally accept. To become more Darwinian than Darwin, biologists had not only to assimilate the idea of natural selection, but to learn to think like Darwinians about a wide variety of different situations.

• •

Marketing Modernism

How does the art world come to define a style of art, and appreciate it as a movement? There are different stories to be told about different styles, but Robert Jensen's *Marketing Modernism in Fin-de-Siècle Europe*, the story of the consolidation of Impressionism, is an interesting and important one.

Key to even the *idea* of Impressionism were people like Paul Durand-Ruel, a Paris dealer whose stable of young artists in the 1870s through the 1890s included Manet, Monet, Pissarro, Renoir, Cézanne, and Degas. He gathered the nucleus of a definable style. Durand-Ruel was a new type of art dealer, trading not only in individual pieces, but in careers. For the Impressionists he performed the invaluable service of arguing, even to the placement of canvases in his gallery, that they were logically descendants of an earlier generation that included Delacroix, Millet, Corot, and Courbet, whose importance had been established by the French Salon. In the Salon system the Academie des Beaux-Arts gave medals at exhibitions, and thereby established the merit and marketability of particular artists. Durand-Ruel claimed a certain disinterestedness in the market by being a promoter of artists and styles of art. He made skilful use of his personal collection to manipulate the tastes of

Honoré Daumier, *L'amateur d'estampes* (1860-63). The solitary image of the print collector is found in several of Daumier's paintings. Of this one, Bruce Laughton, an art historian and biographer of Daumier, writes, "Firmly and competently painted, this small canvas sets a new standard which must itself have been aimed at a new collector's market: the *modest* collector's market."

his clients, and he pioneered the single-artist retrospective exhibition, which educated patrons into an appreciation of contemporary works. He could grumble about losing money on exhibitions, even while they helped to sell expensive paintings, especially to American and German clients flooding into Paris every spring.

Durand-Ruel's strategies were quickly emulated by other art dealers across Europe. And the styles those dealers displayed were consolidated by artists and critics who wanted to demonstrate their own modernity, and to assert their own independence from local academies and established art traditions. Impressionism, then, helped to create the independent modern artist, and a modern art market that provided validation of artistic merit.

Ideas do not spread of their own accord. However innovative individual Impressionists were, there was no necessity that the public accept them as an important artistic movement. That required the determined but quiet work of dealers, artists, and investors. There is no parallel marketing of scientific ideas, but nonetheless they do not take hold without some work to disseminate them. As in the case of art, the value of scientific ideas has to be shown. Value only appears to be obvious once people have worked to make it obvious.

• •

What Happened to Logic in Science?

What have we learned so far about reasoning? Scientists, along with everybody else, learn patterns of reasoning about specific domains. Much of that learning takes place in interaction with other people, though sometimes it is imaginary interaction, as when scientists try to predict the responses of reviewers and critics. Patterns of reasoning are not pure logic, the stuff of the rationalist myth. Not that scientists are illogical or irrational, but logic doesn't carry them very far.

Logic, while very pretty, is by itself quite useless. Logic is self-contained, and doesn't by itself hook onto the material world. It is the science of statements, describing which statements follow strictly from others. Most of the time very little follows logically from what we know, and when we think that it does we learn that most of our knowledge consists of generalizations that don't hold

100 per cent of the time. It takes hard work to connect logic even roughly with the world, to fine-tune categories to match the real and useful divisions of the world. The difficult work of scientific reasoning isn't that of figuring out what is logical, but of figuring out what makes sense. Did it make sense to see organisms in terms of their adaptations, and to see non-adaptations as small exceptions to the rule? Or did it make sense to see other organizing principles as primary, and adaptation as secondary? Were patterns in the fossil record real, or were they artifacts of a poor record and over-eager arrangers?

Reasoning is more challenging than logic. It is difficult because divisions and kinds of things in the natural world are not self-evident, and because recognizable exceptions are all too common. Therefore, it is never easy to know when and how to apply theories. Were the unexplained cases that Darwin faced exceptions, or merely problems to be solved down the road? As exceptions, did they confound the theory, or merely provide special cases, for which other theories needed to be developed? The answers to these questions do not depend on logic, but on a sense of how important these cases are and how likely it is that they will be explained. Every theory is born refuted, but that does not prevent scientists from having strong intuitions about the potential of their theories, using them, and working to explain away the apparent anomalies.

Particular modes of reasoning are specific to particular domains, to combinations of objects and processes. We can see why from the structure of the disagreement between Darwinists and orthogenecists about what constituted an adaptation, and what constituted a pattern in the fossil record: for every domain one of the key things scientists have to learn is what they are looking at. In some cases they can develop precise definitions, but in many more cases they simply have to learn how to perceive the world well. For biologists to think like Darwinians is a skill, and a skill that takes time to learn.

According to the rationalist model or myth of science, what is important about scientific thinking is completely independent of the immediate context of that thinking. If scientific thinking is essentially applied logic it must function universally; i.e., every rational person, when faced with the same ideas and the same

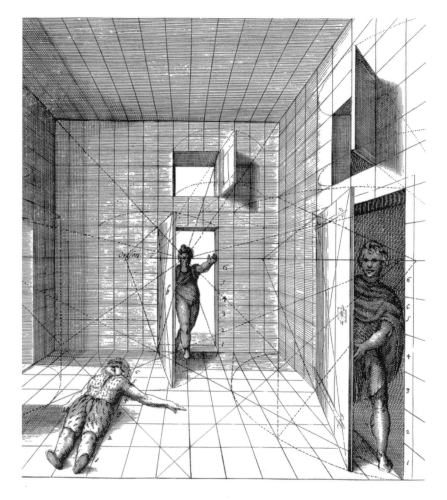

Artists have often used tools and techniques to mechanize and routinize aspects of their work, and sometimes the development of these tools brings on profound changes in artistic styles. Nonetheless, in the terms in which we conceive art, a purely mechanical artist is a contradiction in terms.

evidence, should come to the same conclusions. If this were correct, then we would have a hard time understanding the early resistance to Darwinism within biology. On the rationalist model, it is all too easy to see the nineteenth- and early twentieth-century critics of Darwin, and the many biologists who found natural selection inadequate for their purposes, as illogical and maybe even incomprehensible.

In contrast, on the model of science as art, the difficulties that nineteenth-century scientists had with natural selection look a little more understandable. We can explain those difficulties in terms of the large amount of work needed to develop and expand specific reasoning skills. Put differently, we can understand what it means to "grasp" an insight in terms of something more than simply understanding the words on a page; it requires knowing how to apply the insight. For many people to gain that skill, useful examples have to be found and have to proliferate; and people have to learn how to extend those examples to novel cases, sometimes making genuine theoretical innovations in the process. So, to say that scientific reasoning is a learned skill is just to say once again that science is an art. It takes experience and a practical understanding of important concepts and distinctions. It is an art because a machine could not do it nearly as well as an experienced person can. And it is an art because at least so far in the history of science, there are few last words: the development of reasoning around a topic rarely comes to an end.

That said, the sciences are not much like today's *fine* arts in this respect. True, to be successful a new artistic style has to be explored, criticized, and refined. Then it might last for a number of years. But at least now—many earlier artistic traditions were different—artists and their publics value novelty much more than the sciences do and value consensus much less, so styles are apt to change quickly and dramatically. There can be no myth of the computer for painting, sculpture, dance, or music, because modern artists do not develop traditions for long enough to appear rigid. When the art appears to have been produced mechanically, we say that it has failed.

To summarize, scientific reasoning develops over time. Scientists have to find patterns of reasoning that work for particular types of cases, that depend upon learning the important features of those cases. Then they try to extend those patterns to new types of cases, which requires understanding or discovering what remains the same between the old and new cases. It also requires learning what the important features of the old pattern are, to be able to apply those features in novel contexts. In this way patterns of reasoning may be revised, and even rejected. In all of this, logic is almost never at issue: most people agree about what logic is, even while they may disagree about what is reasonable.

CHAPTER FOUR
• • • • •

Controversies and Consensus

•

*Scientific knowledge is importantly not what individuals believe,
but what communities believe. Communities, however, do not
always arrive at consensus without a struggle. When controversies arise they may rage. During controversies scientists pick apart
each other's evidence, reasoning, and conclusions: almost everything can be called into question. Thus controversies afford
onlookers a glimpse into how uncertain scientific claims can dis-
integrate or solidify into taken-for-granted knowledge.*

Science as a Social Art

From a question of how individual scientists think, we have come
to questions about how scientific communities think. To see science as art implies seeing science in social terms.

Communities build up canons of rationality that apply to
specific domains, learning appropriate ways to think clearly
about certain problems, even defining what constitutes clear
thinking about those problems. This is one aspect of scientific
progress. Science is social knowledge in that what is valued, the
best theories, methods, approaches, and tools are those that scientists agree upon. That amounts to a tautology, and functions
rather like Darwin's principle of natural selection. But just as we
can better understand the living world armed with Darwin's
principle, so we can better understand science with this principle. The things that scientists find most useful, and pass on to
their students and colleagues, survive in the scientific canon. A
claim will be useful if, among other things, it stands a good

◄   Sebastien Le Clerc, "Studying Zoology at the Jardin des Plantes" (1669). As
depicted here the practice of zoology is organized to produce what Le Clerc
called "assured and recognized truth." An established reference work provides
a point of contact for, and perhaps guide to, observations from the dissection.
Those observations are then recorded, making them less ephemeral and more
public.

chance of being noticed and accepted by other scientists. A pattern of reasoning is useful if it helps people to do more research, or to make claims that more colleagues will agree with. And of course knowledge is deemed useful that can be developed in technological directions. A well-adapted idea or skill must fit into a complex environment; therefore, not every idea or skill will survive in science: because existing scientific knowledge has been accepted by most of the community, novelty has to meet many constraints.

Science is also social because scientists are very thoroughly socialized. Years of training result in broad agreement. In this sense, language, problems, patterns of reasoning and conceptual resources all are community resources, reproduced in the individual. Even knowledge is a community resource: though textbooks and journal articles are corrected and reinterpreted frequently in the course of scientific research, they are also in some sense storehouses of knowledge. Textbooks in particular contain what the scientific community considers objective knowledge.

Scientists are heavily interdependent. They use each other's results, methods, tools, ideas, equipment, and skills. Many of these resources are relatively expensive to create anew, and thus they are exchanged, shared, or held in common. As a result, there is a well-developed system of credit, in which researchers acknowledge their use of resources produced by others. This is one way that knowledge is made public and is established in public.

The evaluation of facts depends on the publication and scrutiny of claims, because there is no other marker of a scientific fact than agreement by the community of experts. There is no shortcut to finding out about the facts of nature, not even for experts. Therefore, in an important sense individuals don't create scientific knowledge; they are vital nodes, as a single person may put forward claims, and argue for them, but the transformation of claims into knowledge is done by communities.

Scientific communities, though, do not think with one mind. As we saw in the last chapter, scientific arguments are not always convincing, even arguments that are later seen as strong ones. When bold new ideas are put forward, disagreement is at

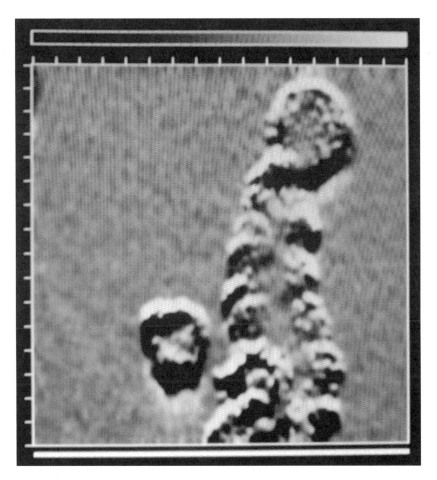

From imagination to visualization: the first photograph of a DNA strand, using the Lawrence Berkeley Laboratory electron microscope. The double-helix structure is already apparent.

least as common as agreement. From the outsider's point of view, controversies are particularly interesting, because they make explicit some of the processes of scientific thinking. In the course of a controversy the disputants raise questions about each other's arguments, and even about each other's skills and integrity. In so doing they show what reasonably can be challenged and what cannot. For this reason we turn to an ongoing controversy, a debate about what molecular biology can contribute to anthropology.

A Conceptual Innovation:
Random Molecular Evolution

As can be seen from opening almost any day's newspaper, molecular biology is changing the fields of genetics and medicine. With increasing knowledge of the structure and processes of genes and their expression, there has been an explosion of interesting questions to be asked. Perhaps the most prominent face of molecular genetics in recent years is the Human Genome Project, the initiative to identify and map every human gene. A complete map holds out promise for medicine, because individual genes sometimes can be linked to diseases and their effects can be studied. With enough knowledge researchers may be able to mimic or counteract the effects of particular genes. Without careful regulation, the map also has the potential to lead to new forms of discrimination, as particular traits become linked to identifiable genes. The Human Genome Project will certainly be opening up exciting and controversial new research possibilities for the foreseeable future.

Another prominent story coming out of the new molecular biology is also subtly altering thinking about evolution. This story provides another illustration of a controversy in science. Though the controversy is not over, it will teach us something about the shape of science when all is said and done.

In 1987, two graduate students, Rebecca Cann and Mark Stoneking, and their supervisor, Allan Wilson, a prominent molecular biologist, published an article in *Nature* entitled "Mitochondrial DNA and Human Evolution." In that article, they presented a striking genetic analysis showing that all humans share a "recent"—dating from only about 200,000 years ago—female ancestor. It didn't take much imagination for people to start calling this ancestor "Eve," and this name has stuck. To make their claims, Cann, Stoneking, and Wilson had made use of an interesting non-Darwinian feature of DNA evolution, the fact that at the level of DNA, evolutionary changes can be treated as though they are random, rather than as adaptations to an environment.

The Random Evolution
of DNA

Despite its distinctive shape, DNA can be thought of as strings of four possible nucleotides: adenine, cytosine, guanine, or thymine (A, C, G, or T). The sequences of the nucleotides act like the letters on this page, forming words and sentences, which are systematically "read" by the cell in its production of innumerable different proteins. Over the generations the sequences of nucleotides change, through accumulated mutations that substitute one for another (an A for a G, for example), through additions or deletions of whole sequences, or through rearrangements. Most of these changes have little or no effect: in our immense genomes large chunks of DNA are filler, as if whole pages and whole chapters of a book were made up of purely random letters. Even parts that are not filler can usually change without creating significant changes on the macroscopic level. Such "selectively neutral" DNA is, therefore, not subject to Darwinian natural selection. But since most DNA is selectively neutral, thinking like a Darwinian about DNA is different from thinking like a Darwinian about whole organisms.

DNA contains traces of its own past, which can be used to reconstruct evolutionary history. We know, for example, that humans, chimpanzees, and gorillas are—in the evolutionary scheme of things—related, which means that the three species share a recent common ancestor. But before looking at their genomes nobody knew whether humans were more closely (that is, more recently) related to chimps or to gorillas, or if chimps and gorillas were more closely related to each other than either is to humans. By looking at comparable sections of DNA in the three species, researchers could answer this question, simply by seeing which of the DNA sections are most similar: consensus now has it that humans and chimpanzees share more DNA than either does with gorillas, and so are more closely related to each other. At the scale of evolutionary change, humans and chimpanzees are like siblings, and gorillas are first cousins. Researchers can also estimate that between four and six million

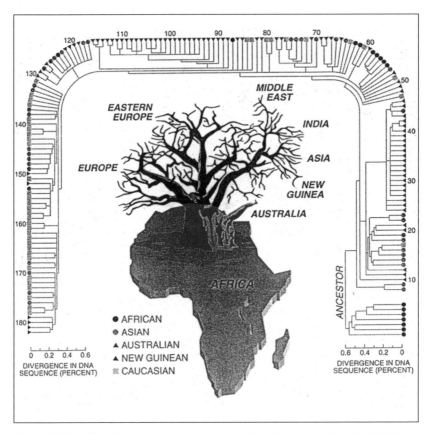

Out of Africa. The disputed phylogenetic tree based on mitochondrial DNA.

years ago humans and chimpanzees shared their most recent ancestor! In the long run, randomness may as well be regular: for example, individual births and deaths do not occur according to schedules, but the editor of a city newspaper can budget roughly the same amount of space each day to the births and deaths columns. If changes to DNA are random then they should accumulate pretty regularly. The difference between two related sections of DNA is a rough measure of the time separating them, forming what is known as a "molecular clock."

The Wilson team used this molecular clock to study human populations. They looked at mitochondrial DNA (mtDNA), which is useful because it is passed on only maternally. The researchers compared mtDNA from 147 different people from different geographical populations, mapping it using 12 different restriction enzymes. Restriction enzymes cut up DNA at particular sites, and by comparing the length of the fragments that they produce in different samples, one can infer information about differences in the restriction sites. The number of differences showed that the 147 people sampled shared a single ancestor between 140,000 and 290,000 years ago, depending on how quickly the mtDNA molecular clock ticks. The mitochondrial DNA of all of the people sampled, from around the world, is descended from one woman who lived sometime in that period. In and of itself, this isn't a particularly interesting conclusion: we know that all people are related, and that means that they share ancestors; despite her misleading name, this "Eve" was simply one of our common ancestors, and not necessarily a special one. However, in combination with other things that we know about human evolution, the figures say a lot. There are widely scattered fossils of archaic humans dating from considerably before 290,000 years ago. Some Chinese fossils are at least 750,000 years old, and many African fossils are considerably older. If everybody around the world can claim Eve as an ancestor, then it looks as though one population of more modern humans replaced all of the archaic ones.

The analysis showed something else that was startling. Cann, Stoneking, and Wilson tried to create possible family trees for their mtDNA samples. The most "parsimonious"—the neatest, in that it required the fewest assumptions—family tree that their computer program produced had only Africans on one side of the oldest branch, strongly suggesting that Africa was the source of modern humans, because all other branches of the tree are branches off an African tree. The molecular clock has plenty to teach us about our history.

• •

The Rhetoric of the Fact

Although we may think of scientific writing as lacking in style, it is better thought of as having tightly defined styles, designed for particular purposes. Consider the article in *Nature* announcing Mitochondrial Eve, by Rebecca Cann, Mark Stoneking, and Allan Wilson, which begins:

> Molecular biology is now a major source of quantitative and objective information about the evolutionary history of the human species. It has provided new insights into our genetic divergence from apes and into the way in which humans are related to one another genetically.

With this the authors announce that the article can be placed in the context of a significant field: molecular studies of human evolution. By referring to so many (14) published articles Cann, Stoneking, and Wilson place the novelty of their work against a background that their readers should accept. Both novelty and conservatism frame every achievement, and therefore make their way into every scientific article. One of the tasks of scientific writing is to create a balance between the two, a building on, reframing, questioning, uncovering, or abstracting from what has come before.

The language here is extremely straightforward. The authors could have chosen more technical language, and could have dispensed with their gentle introduction, which continues for another dozen sentences. Scientists write primarily for their colleagues, though in a generalist journal like *Nature* an article will reach the widest circle of colleagues. But the results were likely to spark a lot of interest, and for that reason Cann, Stoneking, and Wilson wrote so that as many people as possible could understand the problems they were addressing and the results they arrived at. This was an article aimed from the beginning for the spotlight.

Nonetheless, the language becomes more dense and technical further on. When the authors turn to their methods and results, they write: "Each purified mtDNA was subjected to high resolution mapping with 12 restriction enzymes (*Hpa* I, *Ava* II, *Fnu* DII, *Hha* I, *Hpa* II, *Mbo* I, *Taz* I, *Rsa* I, *Hinf* I, *Hae* III, *Alu* I and *Dde* I)." That is a sentence for specialists, and to show the rest of us that the researchers have technical resources on their side. Perhaps the most striking thing about most modern scientific writing is its density: the typical article is packed with technical terms, with difficult-to-read charts or graphs, or with mathematical or statistical formalisms. This article is something of an exception, though density is packed into the corners.

Cann, Stoneking, and Wilson's research on human mtDNA does not lend itself to statistical tests, because it is not clear against what to test their model. Nonetheless, the field demands a gesture in that direction. The authors acknowledge that the human family tree they found to be the simplest, and therefore most likely (reproduced on p. 84), is not the only possible model. However, they say that all of the most likely trees that they examined share the same basic features, including one all-African branch. When they compared their favourite model to a quite different one (the "population-specific tree") "the minimum-length tree requires fewer changes at 22 of the 93 phylogenetically-informative restriction sites than does the population-specific tree, while the latter required fewer changes at four sites ... the minimum-length tree is thus favoured by a score of 22 to 4." Such figures shift the burden of proof, because they show that Cann, Stoneking, and Wilson's model is a potential explanation of the divergent mtDNA samples. Once that has been established, critics have to show that something hasn't been taken into account, or that some other model explains the phenomena better.

The philosopher and anthropologist Bruno Latour says that reading a scientific article is like watching "one player's strokes in a tennis final." In a well-written article, everything fits together to convince readers of the relevant facts, and to convince them that each fact is important. Each sentence, each reference, and each figure is part of the argument. An article like "Mitochondrial DNA and human evolution," though appearing to have been written impersonally and objectively, is simultaneously a passionate argument. We don't normally see the forcefulness of scientific writing because it hides within an objective style. Facts are reported flatly, in a monotone. There is little playfulness, little tolerance of ambiguity, and little suggestion of the human motives that drove and shaped the research. Although there are always deep metaphors in scientific thinking, the surface metaphors are generally of the most trite kind, metaphors on the verge of death. The dry conventionality of scientific writing serves to reinforce its claims to absolute knowledge.

Given what scientific articles are supposed to accomplish, it is no surprise that intense care goes into their writing. Especially if it has multiple authors, an article may be revised again and again. When it is submitted to a journal, reviewers usually make comments that amount to demands for further revisions if the article is to be accepted. And if it is rejected, the authors may want to rewrite it before sending it to another journal. It is not uncommon for a controversial piece to make its way to four or five journals, changing each time. Those changes can affect even the central claims, once reviewers have pointed out limitations or possible problems. Hence, a revision can amount to a re-envisioning of the research. Even scientific writing, then, is essentially social, constructed in anticipation of its critical reception, and in response to actual criticism in the review and revision process.

• •

Adding Complexity: When Really Was Eve Born?

The notion of a recent African Eve immediately provoked controversy. Although almost every anthropologist agrees that Africa was the *original* source for humans, the period 140,000 to 290,000 years ago seems too recent for there to have been a single source for modern humans. Some anthropologists had been arguing that the different populations of humans, spread over the world, had evolved in parallel, with the help of small amounts of gene flow among them. Neanderthal skulls have some of the features of modern Europeans, and some Asian fossils have features of modern Asians. Other anthropologists simply thought that the molecular biological tools were too crude to replace conclusions based on fossil evidence.

When science becomes controversial, the arguments become technical; as a result we could not tell this story without including some of the details. Cann, Stoneking, and Wilson were criticized for using African-Americans as representatives of African genes. They were also criticized on the grounds that they used restriction enzymes, rather than actually sequencing the DNA, which would be more accurate. DNA sequencing became much easier shortly after they did their original work, because of the development of DNA amplifiers called Polymerase Chain Reactions— molecular biology is so young that its methods are changing extremely fast.

More serious was a challenge of the genetic family tree by David Maddison of Harvard University and Alan Templeton of Washington University in St. Louis, Missouri. Maddison and Templeton showed that there were more parsimonious trees than those the original research had come up with, and which did not show African origins. Given the large amount of data, the computer program that had been used was essentially guaranteed to miss good trees, and more runs of the program turned up some of them. The implication was that the original researchers hadn't done a very thorough study. What is more, Maddison argued that there were other interpretations of the originally published tree, interpretations that didn't imply African origins. Clearly, scientists hadn't yet decided what conclusions could be drawn from this sort of molecular data, what patterns of reasoning were good ones. But things were not looking good for the African Eve.

Allan Wilson died in 1991, but scientific controversies are not always centred on individuals. The Wilson Laboratory, in an article written by anthropologist Linda Vigilant and a number of co-workers, conceded that many of the questions were legitimate and some of the criticisms were valid, while maintaining that the conclusions were correct. Students and researchers in the laboratory, utilizing the new DNA sequencing tools, were able to arrive at the same conclusions that had been based on less precise restriction enzymes. They compared sequenced mtDNA from 189 individuals, 121 of whom were native Africans; they were able to show that their earlier use of African-American genes hadn't affected the results. Even before Maddison and Templeton's challenges the Wilson group had changed their method of creating family trees and, using new methods, were

Gustave Doré, *Eve and Adam.*

again able to argue that Eve was African. With their new data they were also able to show that African mtDNA had more variability than European or Asian mtDNA, a conclusion that was drawn independently by the Stanford geneticist Luca Cavalli-Sforza. Vigilant and her co-workers narrowed their estimate of when Eve lived to between 166,000 and 249,000 years ago, though they also found data that placed her at slightly before and slightly after that period.

Templeton, in an all-out attack in 1993, criticized almost every aspect of the research. In particular he questioned the validity of the molecular clock. There is some question about the rate at which the clock ticks, and also about some of the calibrating dates; combined, Templeton argued, these make Eve's dates extremely uncertain. In addition, there is some evidence of continuous gene flow among populations, which would make drawing family trees close to meaningless.

More and more scientists weighed in. Maryellen Ruvolo, a Harvard molecular anthropologist—a designation that could not have existed only 10 years earlier—had been one of the early critics of the Eve research. But in her own study of mitochondrial DNA she roughly confirmed the original dates at which Eve must have lived. Milford Wolpoff, a paleoanthropologist at the University of Michigan, and Alan Thorne of the Australian National University argued that molecular clocks are unreliable in comparison with the fossil record. That is, they challenged the very basis of the research—the regularity of molecular changes. We can see the challenges going deeper and deeper! If building a scientific argument is like building a tower, we can see the critics reaching lower and lower to try to remove supports, hoping to topple as much of it as possible.

In one of the meticulous critiques of the data, Christopher Wills, an anthropologist and geneticist, suggested that molecular clocks are much more complicated than earlier approaches had assumed. Different types of molecular changes occur at different rates, and there may be "hot spots" at which DNA can change very rapidly. Both of these factors affect conclusions about the time since Eve lived, and thus Wills argued that a time frame of 436,000 to 806,000 years is the best estimate. However, for molecular biologists Wills's dates remain outliers, and as a result his criticisms have been frequently ignored. This is an interesting phenomenon: if the molecular biologists agree on Eve's dates, then they are likely simply to ignore methods and arguments that lead people to significantly different dates. Thus, what gets to be considered the best reasoning may depend on what scientists decide the truth is, not vice versa!

Despite her critics, some parts of Eve's story are looking secure. More and more work is being done that follows the lead of the original researchers. Even the prominent critics are starting to accept Eve's dates—but they are questioning her significance. For Eve's supporters, for example for Mark Stoneking, who has continued to be one of the most prominent researchers in this area, the DNA evidence supports what we might call the "Replacement Hypothesis": early humans scattered around the Old World (the Americas were populated later) were more or less replaced in the past 200,000 years by modern humans coming out of Africa. In contrast, Alan Templeton and Milford Wolpoff, who have continued to be among Eve's main opponents, think that she says nothing about the Replacement Hypothesis. They are proponents of what we could call the "Simultaneous Evolution Hypothesis": that the different populations evolved together over the past million years or so, aided by mixing among the populations and natural selection. If this is true, then a common ancestor 200,000 years ago only indicates that there was considerable mixing.

More fronts in this battle keep opening up. Evidence for the Simultaneous Evolution Hypothesis is that older fossils in Europe and Asia bear some resemblances to modern Europeans and Asians. Neanderthal fossils, for example, are thought to have some European features. Unfortunately, Neanderthal DNA extracted from a fossil is quite different from modern human DNA. The common relative of Neanderthals and modern Europeans is more than 500,000 years old. So Neanderthals *were* replaced, and any similarities between them and the people who replaced them is coincidental. Yet another victory for Eve and the Replacement Hypothesis—but not a decisive victory, and Alan Templeton disputes even the victory, holding out for better data.

There have been to date many comparative studies of human DNA, and they have turned up some surprises. It is now abundantly clear that all sections of the genome do not evolve at the same rate; even sections that should be neatly comparable do not. Some of this can be explained in terms of bottlenecks and rapid expansions of population sizes, which produce different effects on the X and Y chromosomes and on quickly mutating and slowly mutating genes. Some of it can be explained in terms of migration patterns: one perhaps unexpected result is that over the very long term women appear to have migrated more than men, possibly because in patriarchal societies they tend to move to where their spouse was born, rather than vice versa.

The story goes on. We cannot recount all of the interventions in the Eve debates. The fact that there are so many, though, illustrates just how difficult it is to cut through to the truth. Although a few things are coming into focus, and support seems to be building in her favour, the scientific jury is still out on Mitochondrial Eve. Almost every piece of evidence can be disputed, and has been, on all sides. For controversial science there is nothing unusual about this situation. It might still take some time for a consensus about human evolution to settle out of all this confusion, as scientists decide what the best tools are, what the facts are, and what the best interpretations are. When that consensus emerges, though, we will have a much better grasp of, among other things, how to think about molecular biology in evolutionary terms—yet another instalment in learning to think like a Darwinian.

Revisiting the Myth of the Computer

Most scientific articles are barely noticed, read only by a few specialists. These may be quietly absorbed into stocks of background knowledge, as possible truths, or potentially useful approaches to a problem, or they may be ignored and forgotten. But when a claim raises a stir we see what scientific knowledge is made of. Suddenly very little can be taken for granted, as even established techniques and pieces of evidence are thrown into question. Suddenly there are many possible interpretations of the data. Suddenly scientific thinking looks not at all computer-like, but just as fraught with uncertainties and differences of opinion as every other type of thinking. It takes an enormous amount of work to address all the doubts and to achieve something like consensus. True consensus rarely is achieved, because there are almost always a few people willing to disagree—and in principle one can always raise more questions if one works hard enough. But eventually debates die down, when it becomes too difficult to raise new questions, and when most people stop taking one of the sides very seriously.

If science involves controversies in which almost everything is up for grabs, where does the myth of the computer come from? One source is the self-presentation of scientists. Scientists typically claim, to each other and to more general publics, that their results come directly from nature. They claim, and indeed believe, that they add nothing to what nature has to say, that they are mere conduits. It is part of the ethos of science to defer to nature. Although there is a sense in which this is correct—scientists discover truths—it is also misleading.

• •

Scientific Writing Has a Style

The first of the two passages below is from a report in the *Philosophical Transactions of the Royal Society* in 1665, and the second from the same journal in 1995.

Some *Obſervations and Experiments upon* May-Dew.

HAT ingenious and inquiſitive Gentieman, Maſter *Thomas Henſhaw*, having had occaſion to make uſe of a great quantity of *May-dew*, did, by ſeveral caſual Eſſayes on that Subjeᒑ, make the following Obſervations and Tryals, and preſent them to the *Royal Society.*

That *Dew* newly gathered and filtred through a clean Linnen cloth, though it be not very clear, is of a yellowiſh Colour, ſomewhat approaching to that of Urine.

That having endevoured to putrefy it by putting ſeveral proportions into Glaſs bodies with blind heads, and ſetting them in ſeveral heats, as of dung, and gentle baths, he quite failed of his intention: for heat, though never ſo gentle, did rather clarify, and preſerve it ſweet, though continued for two months together, then cauſe any putrefaᒑion or ſeparation of parts.

The most usual resistance model used to describe pollutant deposition is the V_d and R_c approach of equations (2.1) and (2.2). While this may adequately describe NH_3 dry deposition to unfertilized ecosystems, as has been discussed, it is less suited to modelling bidirectional fluxes. Wyers *et al.* (1993*a*) have parametrized NH_3 emission/deposition using R_c estimates. However, external criteria were required (e.g. H, light intensity) to impose the required switch from deposition to emission. An alternative to the R_c model is to assume that NH_3 is exchanged solely through stomata with a compensation point, χ_s. In this case the flux is

$$F_g = (\chi_s - \chi\{z\})/(R_a\{z\} + R_b + R_s). \qquad (5.1)$$

However, this model is also limited, since it ignores the parallel deposition to leaf cuticles, controlled by the resistance R_w.

Although they are reports on not totally unrelated topics, they are about as dissimilar as one might imagine. The development of modern scientific styles of writing is part of what supports the myth of the computer: in the second passage there is no recognizable human action, no easy point of entry or dissension.

• •

A second source of the myth is that controversies like the ones we have described produce reasoning that is increasingly difficult to refute, because it is increasingly well justified. Like Dr. Tyrell in *Blade Runner*, experts may deftly respond to questions, justifying their responses curtly with statements of established fact. But, when controversies are raging those statements of fact may be probed, and sometimes undermined by other experts. Controversies die down when challenges become unproductive and difficult, when the work to raise questions becomes too much to undertake given the chance of success. Attacks on and defences of positions refine patterns of reasoning until they are solid in the community's eyes. Scientists, but much more so non-scientists, have little choice but to accept solidified reasoning: those patterns become necessary and define rationality. The social processes in disagreements, then, produce the very rigidities that pass for pure rationality.

Abandoning a rationalist model of scientific thinking does not imply that science is irrational. Nor should it make the romantic genius model any more plausible. Opening rationalism and romanticism to scrutiny should make us look differently on

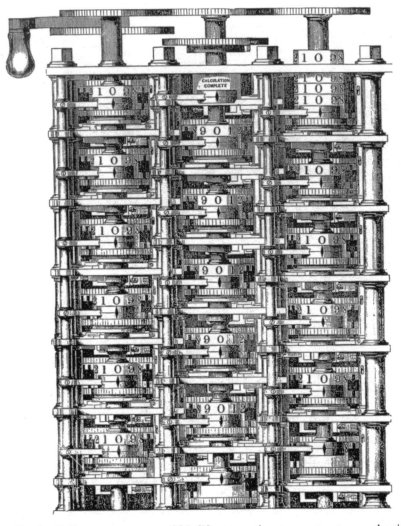

Charles Babbage spent most of his life attempting to construct a mechanical computer. The first attempt, the Difference Engine (1821-1833), was intended to increase dramatically the speed of calculation of entries on navigation tables. Despite investment by the British government the Difference Engine failed, requiring too much precise machining for the amount of financing available. Babbage, however, scaled up his plans, attempting to design a more general-purpose programmable computer, the "Analytical Engine."

science, but it doesn't change the status of scientific knowledge. Scientific knowledge remains the product of carefully reasoned arguments and counter-arguments, close observation, hard work in the laboratory, and more than a little insight. Those things don't guarantee truth, but neither should they incline people to disbelieve.

Of course, the lack of absolute standards, such as those implied in the myth of the computer, sometimes allows for egregious bias. Nineteenth-century anthropology was firm in its commitment to a hierarchy of races, with Europeans at the top—and anthropologists drew support for this hierarchy from every theory of evolution. Until recently reproductive biology imagined eggs to be completely passive in their couplings with sperm for fertilization, reproducing at the microscopic level a Victorian view of active men and passive women. Until there is pressure to change—either from evidence that is hard to assimilate or from social change—scientists may reflect widely held stereotypes.

On the whole, seeing science in social terms does not discredit scientific knowledge. In fact, the social nature of science contributes to scientific objectivity. Which claims, arguments, and methods are accepted is not a *subjective* matter, because success is not based on one individual's opinions and objectivity is the product of minimizing subjectivity. Although that is not quite the form of objectivity envisaged in the rationalist model of science, it is as good as any that real-world science can hope for.

• •

Science in the Public Sphere

The public sphere tends to magnify controversies, and sometimes even creates new ones. Is the globe warming? Newspapers found credible experts to deny it long after mainstream science treated the issue as settled. Does smoking cause cancer? Although most experts accepted sometime between the late 1950s and the late 1960s that it does, tobacco companies have continued to argue that the evidence isn't conclusive.

Courts make this process particularly visible. There, antagonistic interests are not disposed to let mere science settle much. Jonathan Harr's 1995 book *A Civil Action*, later made into a film

Honoré Daumier, *Une Cause Criminelle.*

starring John Travolta, shows what happened to Princeton professor George F. Pinder, "the leading expert on groundwater," who became an expert witness in a large lawsuit. He tried to explain to the court how a factory's illegally dumped toxins could make their way underground for several kilometres, to contaminate a town's drinking water. The opposing lawyers did not let any part of his account rest unchallenged. Each and every premise and claim had to be examined and defended. Even though he was mostly successful, by the time he had been on the stand for several days the list of questions for his account had grown large enough for the jury to develop doubts and confusions. In the end, they did not fully believe Pinder, and half the case of the plaintiffs was lost.

One of the key questions in the OJ Simpson trial, which played daily on television screens across North America, was the reliability of DNA evidence that linked Simpson to the scene of the murders of his ex-wife, Nicole Brown Simpson, and her acquaintance, Ronald Goldman. According to the prosecution, the DNA profile of blood at the crime scene matched that of Simpson at odds of 170 million to one, and the DNA profile of blood on a sock at the foot of Simpson's bed matched that of Nicole Brown Simpson at odds of 9.7 billion to one.

At the time of the Simpson trial, DNA evidence was well on its way to becoming a standard forensic tool. There had been noisy debates about it in scientific journals, but they had mostly died down. Since the prosecution's case against Simpson depended heavily on DNA evidence, the defence—which became known to the public as "The Dream Team" because of its combination of high-powered lawyers—had to challenge it head on. To start with, the defence insisted that there was still a controversy, by finding voices still opposed to forensic DNA evidence. They unearthed discrepancies between estimates of the error rates of DNA evidence. The lawyers found problems with the techniques when they were applied to some minority groups. And probably most importantly, they successfully challenged the competence and integrity of the police forensic scientists: even if DNA tests were generally applicable there was a possibility of contamination, by the sloppy handling of samples or even by the planting of blood on Simpson's sock.

The result is well known. After nine months of trial the jury deliberated for a mere four hours and found OJ Simpson not guilty: the Dream Team had made the DNA evidence controversial, and ultimately unconvincing.

• •

Confronting Nature in Lab and Field

It takes hard work to make successful experimental systems, to design the system, to make all the components line up and do what they are supposed to, and to see the right things. In the end, and because of that work, experimental knowledge is knowledge about the structure of what can be done, what can be made, and what can happen, and not simply about what is.

Experiments Are Hard Work

Every science student knows this, and knows that lab classes can sometimes be exercises in frustration. Reactions don't happen as predicted, scales don't settle down, and even steel balls rolling down chutes may seem to disobey Newton's laws. If we judged by the results of high school and undergraduate experiments—the real results, not just the calculations made by clever students starting from the back of the book—everything we know about nature would be thrown into question. Research does not get easier with age. Those who have been most adept at coming up with the right answer can graduate to harder problems. In experimental science everybody works on something difficult, at least some of the time.

The activity and work involved in experimentation lead us to inquire into the skills that come at a premium in the laboratory. As its etymology indicates, the laboratory is a place of work. It is a place where practical know-how, not just theoretical knowledge, is valued. At the most concrete level, know-how takes the form of technical skill, such as that practiced in performing a certain assay, cloning a cell line, or growing crystals. Often these skills are difficult to master and need to be taught actively. Although there are recipes or written protocols for most laboratory tasks, learning how to perform the task just from the recipe

◄   A water-filled barometer, or Toricellean tube, from the seventeenth century. Mercury-filled tubes were more manageable, able to sit on a tabletop.

In Liebig's Laboratory. Justus Liebig (1803-73) was one of the pioneers of the modern laboratory. Liebig essentially invented the idea of the post-doctoral researcher, inviting young scientists from around the world to work with him in his laboratory. Here that is represented by figures wearing different national dress.

is difficult, and sometimes impossible. Subtle variables can affect how a procedure works, and therefore recipes can only be adequate when the procedure isn't a sensitive one, when, for example, exactly how the test tubes are handled doesn't matter. There is nothing peculiar to science in this: it is difficult to imagine learning to drive a car or learning to paint well from a manual. Skills like these need to be taught by someone saying "try it this

Microphotographic apparatus

way," or "you're not focusing on the right things," or even "your elbow is too high."

As one would expect, then, there is a constant migration of scientists from one lab to another, to learn new skills and apply old ones. Doctoral students learn the techniques of the lab in which they work. They might then get a post-doctoral position somewhere else, where they can both pick up new skills and teach ones that they have already mastered. Sabbaticals and shorter research trips are occasions to retool, to learn new techniques, to learn about new equipment, to learn about new subject matter, and simply to collaborate. Science is rarely solitary for very long.

When novices look at the visual products of the lab, pictures of such things as electrophoresis gels of DNA fragments or radioactively tagged proteins, they don't necessarily see the patterns that experts can see. There is simultaneously too much information, and not enough. Is that row of black smudges a congruence or a coincidence? It depends on the context created by the other smudges. Meanwhile, experts may see the patterns that they're looking for, perhaps after some puzzling and discussion, by making the pictures contain exactly the right amount of information. They have to block out irrelevant material and emphasize relevant material, turning the picture into a diagram in their mind's eye. But expertise isn't something that can simply be turned on and off, and therefore it's tricky to balance observation with openmindedness—finding new distinctions between

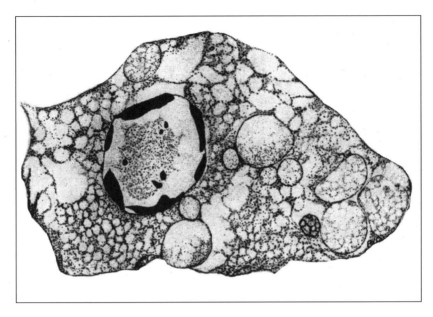

It took years to create easily readable pictures of microscopic images. New dyes, new staining techniques, and new photographic techniques all had to be devised. Even then, published microscopic images are often subtly enhanced to give them a more diagram-like appearance: lines are strengthened, backgrounds are made more uniform, and points of interest are emphasized.

the relevant and the irrelevant—the type of seeing that comes with experience.

The skills involved in constructing and interpreting experiments are not completely different from skills in seeing. Researchers have to know the questions that other scientists will ask of their experiments, and design answers into those experiments or their interpretations of them. All the possibly relevant factors have been taken into account, so that the results can be made to tell as few stories as possible, preferably only one. This is preparation for the sparring matches that take place at scientific conferences or seminars. After a presentation of experimental research, the audience typically interrogates the presenter: Did you try observing when the temperature was lower? Did you act to prevent contamination by ...? Did you take into account the ways in which your organism is peculiar? Are you sure you're not seeing ... instead of ...? If everybody is skilled at this game, they will leave satisfied and confident that any flaws in the experiment, any reasons why the results cannot be taken at face value, have been laid bare. The experimental design and the abilities of the researchers will have been tested, and found either adequate or wanting—as, for that matter, have the abilities of the audience. Presentations are minor rites of passage, but ones that have serious effects on the shape of experiments.

Scientific Observation, Like Artistic Observation, Is a Skill

Anna Brito, a recently graduated medical doctor from Portugal and the subject of June Goodfield's *An Imagined World*, won a fellowship that placed her in an immunology laboratory in England. The lab director, disappointed with the unskilled, non-English-speaking Brito, gave her a task in which she would stay out of everybody else's way: simply to look at a collection of slides of mouse spleen sections, comparing mice that had and had not lost their thymuses.

Partly because Brito had no expectations of what she would see, her patient observation turned up something interesting. In animals that had lost their thymus there was an empty area in the central core of the spleen where lymphocytes were missing. Other cells eventually drifted into the space, but lymphocytes did not. That observation turned out to be extremely interesting to immunologists, and eventually set Brito down a long-term research program of investigating the movement of lymphocytes. It also convinced her of the importance of learning to see, first with an open mind, and then later with what she had trained herself to see: other people had not noticed Brito's empty spaces, and initially did not believe that they were there even when looking at them or at already-published pictures of them.

An Example:
The Colonization of Islands

In 1965, Daniel Simberloff and Edward O. Wilson—the latter now best known for his work on ants and his controversial stance on sociobiology—began a neat experiment that won them an award from the Ecological Society of America. It provides a simple example of the work, skill, and co-operation that go into experimentation, though this particular experiment didn't take place in the laboratory but in the field. Wilson had been the co-creator of a theory of the distribution of species in isolated regions, the "equilibrium theory of island biogeography." The theory was a spare and elegant depiction of the processes of colonization and extinction on islands, constructed largely on the basis of mathematical considerations. Although it has fallen out of favour somewhat, there is no doubt that the equilibrium theory represents an important and fruitful intellectual achievement, and it may yet turn out to be one of the bases for a theoretical integration of ecology.

Wilson and Simberloff's experiment was designed to test the theory's predictions about the colonization of an empty island. The original plan was to watch a normally occurring "natural experiment," in which small islands off Florida are washed bare of life by hurricanes. Hurricanes, however, are capricious, and the idea evolved in a more interventionist direction: Wilson and Simberloff would completely fumigate a number of small mangrove islands in the Florida Keys, eliminating all the insects. They could then watch the recolonization of the islands and keep some other islands as controls.

To do such an invasive study Wilson and Simberloff had to negotiate for permission with the National Park Service. They had to find entomologists and other specialists who would be willing to help classify the organisms seen before and after fumigation—54 specialists helped! And they had to find a method of extermination that killed all the insects but left the mangrove trees alive. The experimenters, not to mention the National Park Service, would not have been pleased to find the islands completely dead; killing the plants would have ruined the comparison between the experimental and the control islands.

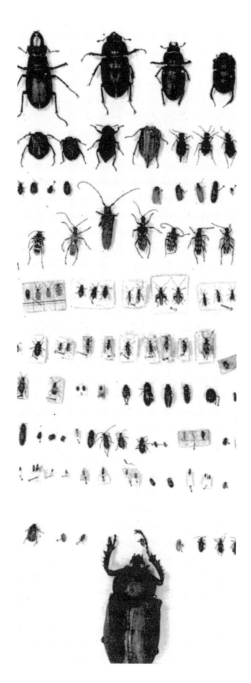

Data take many forms. Charles Darwin, an inveterate bug collector and birder, brought many specimens home with him from his travels.

The researchers began by spraying a few islands with two pesticides, parathion and diazinon. According to Wilson, initially the exterminators wouldn't get out of the boat because of some small sharks in the area; he had to shame them by volunteering to stand guard in the water with a paddle to shoo the sharks away. A few days later Simberloff returned to see the damage, and found that some larvae had survived, buried deep in the wood. These larvae would have affected the results by being left on the island and ready to repopulate it. A simple solution to this problem could have been to use longer-lived insecticide, but that would have the disadvantage of killing some of the new colonists, the ones that Simberloff and Wilson wanted to count. They had to come up with a better fumigation technique.

Their solution was to erect scaffolding over each island and cover it with a tent so that a poison gas could be applied that would get deep into cracks and crevices. Erecting a tent over even a small island represents no small feat of organization, as the artist Christo would later discover in his wrapping of buildings and coastlines. Luckily, the hired exterminator had become intrigued by the project, and after a couple of tries came up with a method of efficiently wrapping the islands. Meanwhile, Simberloff tested out a number of different poisons and dosages and found a concentration of one that would kill even cockroach eggs without substantially damaging the vegetation, and probably would do no long-lasting damage to the area. Even then, Simberloff and Wilson learned to fumigate at night, so that the increased temperatures under the tent didn't harm the mangrove trees: they killed the upper canopy of the first two islands by spraying during the day.

Over the years following fumigation, Simberloff censused the islands for arthropods, turning over leaves, breaking dry twigs, and examining crevices; for the first nine months he was systematically bug-hunting for 12 days out of 18. Out of the record of all the species he found, Simberloff and Wilson were able to create a picture of the recolonization of the islands, a picture that roughly matched the predictions of the theory they were testing.

In retrospect, the results weren't so straightforward. Simberloff later became one of the most vocal critics of the theory and questioned whether he had been recording real "coloniza-tions" or just the routine movement of insects from one location to another. He wondered how many species had been missed between censuses. He even questioned whether these little man-grove islets could be good stand-ins for the real islands that the theory was supposed to describe. None of these issues is extraor-dinary. The experiment had been thought through at the time and carried out with panache: Wilson and Simberloff were fully deserving of the award they won for the research. But method-ological choices have to be made, and data have to be interpret-ed. As we saw in the last chapter, with perseverance any set of choices and any interpretation can be challenged. Therefore, the only crucial experiments are ones that enough of the relevant community decide to *call* crucial—the ones that they decide not to question indefinitely.

Experiments Have Lives of Their Own

There is an interesting contrast between the value of an experi-ment as an experiment and its value as a test of a theory. While many experiments are framed as tests of a grand theory, in most fields there also is tremendous emphasis on simply getting "good data." In fact, scientists can become obsessed with good data, the sought-after reward for a hard week in the lab. Good data or clean results often find their way to publication, even if they aren't neat tests of a theory. Experiments can explore small hypotheses or details; and many researchers have disdain for grand theories, which they see as so vague that they don't con-strain possibilities that people working "on the ground" are con-cerned with. Or, experiments can have value in and of them-selves, and getting good data is another way of saying that the experimental system is working well, which it might easily not be. It is difficult enough to get systems to work properly, and to cre-ate confidence that they are working, that simply "getting the bugs out" can be a highly valued achievement: it creates a tool that can be used by other people. Nobel prizes have been won for such achievements.

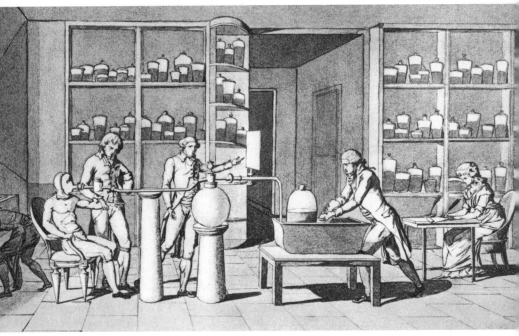

Mme. Lavoisier's sketch of her husband conducting experiments on respiration, she taking notes.

The Experimental World Is an Artificial One

Heraclitus said, "Nature loves to hide." But that is a theorist's aphorism: for experimenters the central problem is the unhelpfulness of the material. "Nature is capricious and stubborn" might be the experimentalist's preferred aphorism. The laboratory is a partial answer to the problem, because the laboratory provides a controlled environment where objects and processes are made more uniform and manageable.

Outside of the laboratory most substances occur in mixtures, so the first step after bringing something into the laboratory is almost always purification. Whether chemicals, metals, or bacteria, substances in their pure form behave much more consistently than they do in their mixed, diluted, and complex natural form. There is, therefore, a significant business for scientific supply companies who provide high-quality inputs for experimentation. Those companies make a business of guaranteeing that their products are exactly what they claim to be, 99 times out of 100, or to one part in 1,000, or however else purity is measured in the particular case.

● ●

How To Get Pure Hormones

When the endocrinologist Andrew Schally wanted to find out the structure of thyrotropin-releasing hormone (TRH), his largest task was to gather enough pure TRH to be able to analyze it. In animal bodies it is found in only minute quantities, far smaller than can provide starting points for analysis. Schally had to come up with a source for "very large" quantities of TRH: a few milligrams. He did this by approaching the meat-packing company Oscar Mayer and asking it to donate and ship the pig hypothalami that it processed to his laboratory in New Orleans. More than one million pig hypothalami came on ice by railway car. Meanwhile, his competitor, Roger Guillemin, with whom Schally shared the Nobel Prize for this research, was having tons of sheep's brains shipped from Chicago to Houston.

Linus Pauling's Nobel Prize for Chemistry.

Schally extracted the substance thought to be TRH, purified the extracts, and then was able to start on the analysis of his milligrams of material. Even then there was too little TRH to do standard chemical analysis, so Schally approached the problem from two directions at once: finding out as much as he could about TRH's structure, and at the same time trying to synthesize molecules that might be TRH, to find out if they had the same properties.

•••

Slightly different from purification, but equally essential to experimental practice, is standardization. To compare results from one laboratory to another, or even to compare results between samples within a laboratory, everybody has to be dealing with the same material, and treating it in roughly the same way. Laboratory mice, for example, are common in laboratories, used for everything from cancer research to studies of behaviour. They were developed as a research tool by the Jackson Laboratory in Maine in the 1920s; the Jackson Laboratory supported and still supports its research by selling the standardized mice to other laboratories. A mouse line is created by mating siblings to each other generation after generation, until the mice are essentially clones of one another. The line can then acquire a label, such as "Dilute Brown Agouti YBrdPA," by which researchers can identify it, knowing that for practical purposes every individual of that line is genetically identical to every other of the same sex.

Measurements are most useful if standardized, as are tools for measurement and manipulation. For example, in the island fumigation experiments, Simberloff became his own standardized measurement device, by bug-hunting on each mangrove island for 14 hours every 18 days, so that his results from one island to another were comparable. The more that researchers and their tools act in machine-like fashion, producing predictable responses to expected stimuli, the more objective they are in the sense that standardization displaces their subjectivity. This is important enough that objectivity is often valued more than accuracy, more than detailed descriptions. The fine arts obviously do not value this sort of objectivity at all.

In addition to purification and standardization, the materials of the laboratory can be enhanced compared with their naturally occurring form. DuPont's "OncoMouse" is a line of mice genetically altered to develop cancer reliably. OncoMouse is a tremendous tool because it removes one source of uncertainty: for these unfortunate mice the cancerous state is the normal state, not an exceptional one. For other biomedical research it is possible to buy mice that have been raised entirely in a sterile environment and are therefore free of particular germs. Such mice live their early lives behind airlocks, and when they are ready to become research subjects they are shipped in sterile containers to the labs needing them. There they can be exposed to certain diseases and treatments in the confidence that they have developed no immunities to those diseases and aren't infected with other diseases that might confuse the results.

Inside the laboratory researchers rarely are content merely to observe the objects in which they are interested. Rather, objects are placed in situations in which they display hidden facets or properties. This process is carried to an extreme in

Probably the earliest operating accelerator: Lawrence's four-inch copper-encased cyclotron. The electromagnetic poles at bottom right and the next clockwise to the left alternate their charge to keep the hydrogen protons moving. A similar system is used today in the charged-particle accelerator on the right.

the largest of today's experiments, those in high-energy physics, which are mind-boggling in complexity even while, in fact *precisely because*, they attempt to study the most pared-down of phenomena. The aim in these experiments is to force particles to collide at very high velocities, in order to watch the fireworks that result. The Stanford Linear Accelerator pushes electrons and positrons along an absolutely straight underground tube two miles long, giving them an extra push with high-frequency radio waves every 40 feet. At the end of a tunnel sits a detector, an instrument that has material for the accelerated particles to collide with, and various pieces of apparatus for

recording what happens when they do. Once the accelerator is up and running, experimenters can pretty much ignore it and focus on particular detectors. But detectors themselves are massive pieces of equipment of enormous complexity. A detector can be the size of a house, is the product of a collaboration involving hundreds or thousands of people, and costs millions of dollars. For the equipment to work, every part of it has to be independently modelled and tested. A small army of scientists, engineers, and technicians continually scrutinizes the machine to make sure that it is doing exactly what people hope it should be doing.

Accelerators produce phenomena that can't be found in nature: large numbers of high-energy collisions of a consistent type. In the volume of these collisions can be seen the patterns that physical theory predicts. But volume has another positive consequence: in recent decades what has most interested researchers in high-energy physics is not the common products of a collision, but the very rare event that occurs one time in a million. The seventeenth-century philosopher and scientist Francis Bacon said, "For like as a man's disposition is never well known or proved till he be crossed, nor Proteus ever changed shapes till he was straightened and held fast, so nature exhibits herself more clearly under the trials and vexations of art than when left to herself." High-energy physicists would have to agree, for those rare events of collision are thought to be particularly revealing about the fundamental properties of matter, and are expected to motivate the next theoretical developments.

• •

The Unnatural Nature
of Experimentation

In his classic air-pump experiments in the seventeenth century, Robert Boyle placed first one thing and then another under the glass of his pump to see how each behaved as his assistants pumped out the air. Candles, small animals, and barometers all were observed as air was removed from the chamber. The typical barometer of the day was a long, thin test tube of mercury inverted into a bowl of mercury. With practice the column of mercury could be made to remain in the tube, settling some 30 inches above the level of the liquid in the bowl. There were two hypotheses as to why the mercury was held up: it was either kept up because "nature abhors a vacuum" or by the pressure of the atmosphere.

Outside of Boyle's laboratory, there were only complex arguments from first principles about the two hypotheses. But inside the chamber of his pump, Boyle had a way of manipulating the atmospheric pressure, if it existed. He found that pumping out the air lowered the height of the mercury, and therefore could argue that the mercury was held up by atmospheric pressure rather than by the vacuum. By placing the barometer in an unnatural environment Boyle could make claims about its nature, claims that couldn't be made simply by watching the normal course of events. Boyle's scientific descendants continue to employ this strategy daily.

● ●

What Are Scientific Experiments About?

In a certain sense, then, nature is checked at the door of the laboratory. In the field, nature's capriciousness and stubbornness always threaten to confound experimental results. To deal with the uncertainty, unpredictability, and messiness of the natural world, experimenters create the most unnatural environment they can by purifying, sterilizing, standardizing, and enhancing their subjects and tools. And inside that environment researchers combine substances in odd ways, pushing those substances to their limits and beyond. These manipulations allow experimenters to arrive at clean and neat results, the sort of results that theoreticians can build on. At the same time they encounter the recurring question: Are the products of the laboratory representative of nature, or are they artifacts, unintentional by-products of laboratory manipulations?

Before the seventeenth century, science—or natural philosophy, as it was then called—assumed the task of accounting for naturally occurring regularities, the sort of thing that all could observe if they took the time: the tides, the movements of planets, the growth and generation of plants, and so on. Laboratory phenomena were used as illustrations, but in general their artificiality did not make them legitimate objects of interest. When seventeenth-century experimenters tried to elevate laboratory phenomena to a higher status, as prompts to and final tests of theory, critics objected that what they were studying was unnatural and foreign to natural philosophy. Experimenters eventually won the dispute, but not because they had a knock-down argument in favour of the laboratory. Rather, scientists simply decided that artificial phenomena have to be legitimate because they are such valuable sources of knowledge, regardless of whether or not that knowledge is about nature in its untouched state. In effect, scientists redrew the boundaries of "nature," to include anything that could be described in scientific terms. Today there is much more to "nature" than there once was.

The controversy never quite goes away, though. Every field has to draw a line between the natural and the artifactual, a line that depends on what is being studied, and on agreement about the legitimacy of those phenomena as objects of study. One of the reasons Wilson and Simberloff's island fumigation experiments were so well respected is that they took place in the field. Even then, Simberloff later questioned whether the little mangrove islands he and Wilson had manipulated were representative of the larger islands that were of more compelling interest. For ecologists, where the line between the natural and the artifactual lies is still an issue perhaps because ecological phenomena are so central to ordinary ideas of nature.

Among other things, then, laboratory results show researchers what objects can be made to do and in which circumstances they can be made to do it. Perhaps this is why science can so fruitfully feed into technology: work in the laboratory is already proto-technology. Of course, the artificiality of the laboratory suggests a problem for turning science into technology, because what works under rarified scientific conditions may not work in the messier conditions of the outside world. Interestingly, technologists usually respond to this challenge by working as well to make the outside world more like the laboratory: building temperatures are controlled, the terrain is flattened or sculpted, people are trained to behave more predictably, and so on.

Theoretical work in science aims not simply to copy nature, as we saw in Chapter Two. Typically, experimental work is held up as anchoring theory to reality. But, while experimental science rests in reality, it is no mere copy of nature, either. If anything, it is more profoundly a human creation than is theory; it is hands-on creative work leading to the construction of beautiful and revealing things. The material and intellectual products of the lab are valued for the cleanliness of their lines and the definition of their contours. Because of this, experiments can serve as a basis for theorizing and as models for action outside the lab.

CHAPTER SIX

Doing Science in the Real World

•

As science grows or moves into new environments, the entwining of science and the social worlds around it becomes more obvious and more complex—so much so as to call into question the boundaries between them. Like other types of work, scientific research is not isolated from the things it touches, but is connected to networks of institutions and changes with them. In fact, the more effective and successful science is, the more connections there are to other things. It requires a balancing act to maintain science's separateness and autonomy while keeping it well-funded and relevant.

Is Science the Same If It's for Profit?

It is May 7, 1991. Joshua Boger, the founder and president of Vertex Pharmaceuticals Incorporated, is facing some serious problems. In a few days the journal *Science* will publish an article on the structure of a significant molecule, FKBP-12, by Stuart Schreiber, a Harvard professor of chemistry. FKBP-12 is probably the key to a new immunosuppressant, and is the only important project that Vertex is working on. Boger has already partially mitigated the damage by convincing the journal *Nature* to publish a Vertex author's article also describing FKBP-12's structure. However, he had to promise that a second article, confirming the first by a different method, would be coming shortly, and one of his X-ray crystallographers has just failed to deliver. Boger now has to figure out how to get his key researchers to work together—

◄ Von Edelfeldt, *Louis Pasteur* (1885). Louis Pasteur is undoubtedly a heroic figure of nineteenth-century science. He is also an emblem of a style of science that simultaneously addresses fundamental problems and aims at application. Pasteur's success in his research on immunization and infectious diseases earned him national fame—there is perhaps a Rue Pasteur in every sizeable town in France—and worldwide recognition.

and they are at each other's throats, near nervous breakdowns from the pressure—to salvage the effort and come up with something that's good enough to pass the reviewers. Worse still, the first Vertex article will appear a week after Schreiber's, and while that counts as a tie in the world of science, the second article won't create any publicity in the business world, which Vertex desperately needs. The stock market is buoyant about biotechnology start-ups, and only a week ago Boger decided to take Vertex public. If Boger could issue an announcement of Vertex's results this week, then *The Wall Street Journal* likely will publish a story that gives Vertex and Schreiber equal credit. But U.S. government rules prohibit publicity designed to increase the value of new shares, and *Nature* prohibits prepublication announcements of results, so making an announcement is going to be difficult. Boger has to convince his lawyers that an announcement is legal, and has to convince *Nature* to waive its normal prepublication rule. Both turn out to be easier than he expected. "This is the value of working in hot areas," one researcher tells another. "*Science* and *Nature* have no idea whether any of these structures are right, but they'd obviously rather publish shitty structures of important proteins than exquisite structures of boring proteins."

The Vertex story, compellingly told by Barry Werth in his book *Billion-Dollar Molecule,* shows something of the frenetic pace of a biotechnology company in its early days. To get investors, Vertex has to come up with some results, but there can be no results before there is money to pay for researchers and equipment. To escape this vicious circle Boger has to spin stories to risk-taking investors, and then convince his researchers to work flat out to achieve some successes; only then can he hope to attract the funds the company really needs to create some new drugs. In the short run Vertex can't afford to be beaten to the punch on anything, so every scientific decision is simultaneously a business decision and vice versa. The research has to satisfy joint scientific and business criteria, leading to the desired results, answering good questions, and looking like the sort of work that will pay off in terms of profits. Disentangling anything like pure science from this tight knot is a nearly hopeless enterprise.

Sébastien Le Clerc, *Louis XIV visiting the Académie des Sciences in 1671.* Royal patronage turned into government patronage and has had subtle and not-so-subtle effects on the shape of science.

Stories like these lead many people to look at commercially driven science with distaste. If we think, however, that market-driven science is *essentially* different from pure science, we are mistaken. Though they are more visible and more exaggerated, the pressures faced by a company like Vertex have their analogues in pressures faced by academic scientists. Academic scientists have to develop track records so that they can count on some research grants down the road. They have to design research programs that are likely to lead to significant and novel results, which may mean subtly avoiding competition with other scientists, even while sticking to questions recognized as important by their disciplines. They sometimes become involved in races for priority and have to plan their moves around those races. Even when there is little or no commercial value to their work, academic scientists, like industrial scientists, sometimes are secretive to prevent competitors from gaining an advantage. And they often have to deal with untidy collaborations and conflicts among egos. In short, there is no scientific research that stands wholly apart from issues we normally consider extra-scientific. As universities, especially in North America, have become home to innumerable start-up companies, academic scientists increasingly are aware that their work may have commercial value, which accentuates all these issues. Nonetheless, "pure science" remains a useful abstraction, rather like a "frictionless plane" or a "genetically determined trait."

Even Pure Science Depends on Impurity

Modern scientists have a tremendous amount of autonomy to distribute and use funds, to evaluate others' careers, and to evaluate others' claims and authority. Few other groups have control over their work to the extent that groups of scientists do. For example, governments, which provide most of the funding for scientific research worldwide, distribute money to science in ways that they would never consider for most other parts of their budgets. After allocating funds to broad areas of research, government agencies allow the scientists themselves to decide who merits how much money for which projects. Experts in the relevant fields read the grant applications and evaluate them. Then, the recipients of the grants are not expected to do more than pursue their projects in good faith

Nicola Testa was the first to propose a practical way to transport electricity by using alternating currents and transforming high-voltage/low-current into low-voltage/high-current power.

and make annual reports. Government funding for the arts is in some cases similar, but involves very much smaller amounts of money and is more tightly controlled. Imagine, however, welfare recipients being given the same freedom to allocate welfare funds, or road contractors being expected only to *try* to complete the roads they bid for.

Key to their privilege is scientists' claim to have almost exclusive ability to judge the merit of their work. Most people feel qualified to judge art and athletic achievement, and even the needs of welfare recipients and the quality of roads, but few people believe that they can judge the merits of scientific research. Everybody is a critic, but when it comes to science, to be a critic takes too much expertise for most people. Thus when articles are submitted to scientific journals, even editors generally defer to the judgements of independent reviewers, who are experts in the relevant fields.

Science is a particularly encompassing social world. For most inhabitants of the world of science, most of the time, research is the main goal, and rewards in the form of recognition by peers are more important than rewards in the form of money or power. Publishing in prestigious journals, for example, is often more important than earning more, because it represents approval by peers rather than by administrators.

Scientific research, then, takes place in social worlds that have some measure of autonomy and some measure of isolation from other worlds. Scientists draw sharp boundaries around the core of science and put much higher value on what is inside than what is outside. For example, they distinguish time spent "doing science" from the rest of their working time, when they do everything else they need to. Researchers tend to value "pure" much above "applied" science, and tend to assimilate their own work to the former rather than to the latter. Drawing sharp and idealistic boundaries like these helps to maintain the integrity and autonomy of science.

No social world, however, is completely isolated, and successful social institutions need to connect with outside people and institutions. A completely isolated social world becomes a cult, easily ignored by everybody outside it. Science cannot afford isolation, because, among other things, research is expensive. There can be no science without constant infusions of resources. To obtain funding, scientists turn to governments, to industry, to foundations. They have to offer at least the potential of something in return: useful knowledge, expertise, a product of economic value, or a prestigious intellectual achievement. Even to pursue pure science, researchers have to engage the world around them. Engagement, however, poses a constant threat to autonomy. Science in the real world requires a constant balancing act between connection and isolation.

• •

Contemporary Women in Science

In 1988, the average salary of a woman in science was about three-quarters that of a man, if each had 10 years' experience.

Of women Ph.D.s in the natural sciences employed in U.S. universities, only 18 per cent were full professors, versus 46 per cent for men—statistics that are comparable with ones for other high-status professions. It is clear that these figures represent tremendous improvements in the status of women from a generation earlier. But considering that science prides itself on being a purely intellectual endeavour, recognizing and rewarding only high-quality work, they raise questions.

The stereotypical scientist is still male. That stereotype can make it more difficult to see a bright young woman rather than a bright young man as a potentially successful scientist. Women frequently observe that they are subjected to more scrutiny by their colleagues than are men, and that sometimes their contributions are devalued. This can have a substantial impact on opportunities and might help to explain, for example, why of 134 male mathematicians who applied for prestigious fellowships in 1990, 21 won them, but of 56 women who applied, only one came out a winner.

Stories of overt and aggressive cases of discrimination and harassment abound, but more common are minor irritations and hidden disadvantages. On the whole, women rely more on published work than do men, because they aren't as often part of as many informal networks. Women are less likely to be involved in collaborative research, an important route to publication—scientific publications have on average nearly three authors. Mary Frank Fox says, "Although women have moved into science, they are not of the community of science. More often than men, they remain outside the heated discussions, inner cadres, and social networks in which scientific ideas are aired, exchanged, and evaluated."

Even the smallest and most subtle forms of discrimination can have big effects when these are magnified over the course of a career. One study of University of Michigan scientists showed that newly appointed men received more research support than women, including bigger initial grants and better facilities. This was echoed by a recent MIT study that found senior women scientists working in smaller laboratories and losing larger portions of their grants to the university. Such differences can have large long-term effects, even if they have only small short-term ones. Because success begets success, even the most minor advantages and disadvantages can become magnified. The more a researcher has accomplished the more opportunities—for research grants, publications, prizes, and even information—he or she will have. Thus, while women are narrowing the gap, as long as there is a gap there is room for progress.

• •

Publication Pressures Shape Science

We saw in earlier chapters some of the work that goes into a scientific article: even after all the hands-on research is done authors may revise their writing many times in response to real or anticipated criticism. This effort is for unpaid writing— publication is so important to researchers that writers of journal articles are never paid, and may even have to pay a small fee per page! Why do they go to all this trouble? Because articles are attempts to establish facts, and scientists are judged in terms of their contributions to knowledge.

When scientists apply for research grants, past publications are one of the important factors in judging whether or not they are likely to succeed in the projects they propose. Even without paying attention to their content, the number of publications is a concrete measure of a researcher's track record, and thus increases the chances of winning research grants. In turn, funding is essential to almost every field of scientific research, so winning grants has a direct impact on the quantity and importance of the research that a scientist will be able to do, and even whether he or she will continue doing research at all and continue to publish. As in most human endeavours, success breeds success.

Not only numbers count. When scientists publish articles that are seen as particularly valuable contributions, they gain visibility and opportunities for high-level interactions with other scientists. They are pulled towards the core of their disciplines. This means that more people are likely to read their work and pay attention to it, so their future research will have more impact.

Published articles, then, become indicators of productivity and indicators of importance. But more than that, publication is deeply woven into the fabric of scientific culture, meaning that in order to be a "productive research scientist" one has to publish. Some of the pressure to publish comes from external sources, such as the need to get jobs or grants, but just as much comes from more internal sources like the desire to be esteemed by colleagues or simply to fit the image of a scientist. Even a scientist's self-evaluation depends in part on public evaluation.

With so much riding on publication, its demands often shape the research itself. This claim need not be seen in a negative light: it means, among other things, that when research is being done the researchers are already thinking about how their peers are going to read the articles they plan to write. Are they addressing the right sort of question, the sort of question that other people would want to see addressed? Is the work on a scale that will lead easily to publication? Are there different angles, so that a single study can be written as multiple articles? What will reviewers likely complain about? What will readers find useful? Scientists don't simply ask a question that interests them, and then go on to answer it, but usually think of themselves as part of a community from the beginning of the research process. Scientific research is a social activity from the beginning.

From Associations to Institutions

The time when individuals, unconnected to any institution, could routinely make significant contributions to scientific knowledge now seems to be past. Of course there are exceptions, people who combine a passion for some subject, the time to assimilate a field, and possibly the resources to conduct some study. But they are exceptions, and the reasons for this should be obvious: over time, the burgeoning of science increased the total amount of knowledge, so that to make an argument and to claim authority required more and more background. As instruments

Benjamin Franklin's famous experiments on lightning.

have become finer, techniques more precise, and standards of data higher, it has become more expensive to do empirical studies. Therefore, researchers need funding, and need the prestige and stability that belonging to some institution brings to get that funding. Two hundred years ago people like Benjamin Franklin and Thomas Jefferson could do research alone simply by flying a kite or collecting reports of travellers. Since then, scientific research has increasingly become a specialized activity, pursued by people in industry, universities, and institutes. And though important work is done by technicians and graduate students, people with doctorates are the key nodes in scientific networks, writing grant applications, directing lines of research, writing articles, validating knowledge, and receiving the lion's share of the credit. (Incidentally, that is another demonstration of the strange place of creativity in science. Most of the credit goes to people who claim to be the creative forces behind scientific work, not the people who simply do the research. If scientists were like computers, all work would be of equal value.)

Modern science started to get off the ground in the 1600s, with a rash of new theories, instruments, institutions, and interest in the workings of the natural world. Since then, its size has been increasing at a tremendous pace. Derek de Solla Price, one of the most prolific measurers of science, has estimated that the number of scientific journals and the membership of scientific institutions has doubled every 15 years for about the last 300 years. By other measures the doubling period has been closer to 10 years. Perhaps more striking is his statement that "using any reasonable definition of a scientist, we can say that 80 to 90 per cent of all the scientists that have ever lived are alive now." Part of what makes science so immediate is that the vast majority of all the work in any given field has been done by people who are still alive.

With this prodigious growth has come the development of innumerable institutions to deal with science. Communities of a few dozen can govern themselves informally, but larger communities need formal rules and formal institutions. A network of researchers 200 years ago would have been small enough that almost everybody in it would have known not to put too much credence in the reports of some unreliable Dr. X. But as networks increased in size, more and more formal restrictions would have to be placed on who could contribute to knowledge. For this reason the Ph.D. has become almost universally required for research science, and gaining a Ph.D. has become an increasingly standardized process. Similar changes have taken place in every aspect of science: formal procedures for accepting an article to a journal have become more common and more standardized, as have formal procedures for funding. The result of this formal requirement is the reopening of space for informality. Socialization within each field is so thorough that members of the field can interact in the context of considerable shared expectations and beliefs. Meanwhile, growth has created bodies that advocate for science and bodies that regulate it—no activity that draws as much resources as does modern science can be expected to be free of such agencies. And as more resources are needed for research, governments and industry are increasingly able to demand that their goals be met when they fund research. So although one of the attractions of a scientific career is the freedom that it affords, as science grows and resources become tight, people who choose that career have had to accept increasing numbers of institutions that encroach upon that freedom.

• •

Big Science

At the turn of the century, C.T.R. Wilson, a physicist interested in meteorology, spent 15 years developing a tabletop apparatus that would mimic cloud formation. He eventually discovered that clouds would form around ionized particles, which could be made to appear in the path of X-rays and gamma rays. Wilson's device became known as a cloud chamber, a key detector of elementary particles for physics in the early part of the twentieth century. Although Wilson drew upon many people's ideas and expertise, he essentially worked alone. His main concerns were with how to make his apparatus work, and with understanding the relation between charged particles and clouds.

Fifty years later, Luis Alvarez built the world's first large bubble chamber, a direct descendant of Wilson's cloud chamber. Alvarez needed not only to deal with cantankerous materials—

including quantities of highly dangerous liquid hydrogen, radiation shields, tubing and gaskets capable of handling high pressures, and large power supplies—but he also had to organize and mobilize many people. He needed to convince government agencies and university sponsors that his project was worth funding to the tune of millions of dollars; he had to manage complex collaborations between physicists and engineers who had very different styles of working; and he had to employ and train a number of scanners, mostly women, to examine for signs

of interesting results the millions of photographs coming out of his machine. In short, by increasing the scale of his instrument, Alvarez had to engage in social and political work quite different from that of his predecessor Wilson.

Luis Alvarez set out to do "big science," just as someone like Christo sets out to do "big art." Christo and his partner Jeanne-Claude's spectacular work, "Wrapped Reichstag," for example, took 25 years from initial conception to completion. The final product, the German Reichstag in Berlin completely draped and covered with white cloth, was wonderfully simple and monumental. But the steps towards that final product were anything but simple. Detailed engineering plans had to be drawn up. The German parliament had to vote in favour of the project, which meant that key government figures and many individual members of the parliament had to be brought onside; Christo and Jeanne-Claude lobbied a staggering 352 members on the project. Funds had to be raised on a large scale; 100,000 square metres of specialized fabric had to be woven. Engineers and hundreds of construction workers had to be hired and organized. In short, the art work was as much the organization and planning as the final product. Big art, like big science, is often different in its very nature from projects of a more individual scale.

• •

Dollars and Cents

Derek de Solla Price made his estimates of the doubling time of science in the early 1960s, and the pattern has not quite continued unabated since then—growth that rapid had to come to an end before it overwhelmed everything else. Yet the size of science has by no means levelled off. By one estimate, in 1965 the number of scientists and engineers engaged in research and development in the United States was 495,000, but in 1988 that number had grown to 950,000. Most other developed countries show similar patterns. In the U.S. total funding for research and development has been increasing steadily: between 1980 and 1991 it increased by about 50 per cent, to $152 billion. A considerable portion of that increase was due to increased military spending in the early 1980s. And the relative portions of government and industrial funding have changed substantially, with industry surpassing government in the 1980s. Those facts are important for the type of science produced, because industry and the military support less basic research—research not geared to commercial development—than does the rest of government. Despite this, the growth in dollars spent on university-based research and development averaged 5.5 per cent in the 1980s. The situation is roughly similar in other major industrial countries, such as Canada, France, and Britain, though generally not as rosy.

From the point of view of scientists competing for research dollars, the increased funding has seemed not nearly enough. There are increasing numbers of scientists competing for that money. The slow shift from government-dominated funding to industry-dominated funding has meant that many scientists have had either to develop more entrepreneurial skills or to accept positions with less freedom than previous generations were used

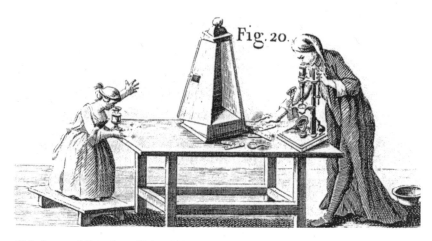

"Mother and Daughter Using Microscope" from Jean-Antoine Nollet's popular five-volume *Lessons on Experimental Philosophy* (1748). This atypical representation of women exploring nature was intended to increase the popularity of science and of Nollet's *Lessons*.

to. A tightening of funds may also have contributed to an increasing obsession over work. Daniel Koshland says, "The work habits of scientists are addictive, leaving their spouses in tears, their children pleading, 'Come home, Mommy (or Daddy),' and involve long hours in hostile instrument laboratories or cold rooms, exposed to noxious gases and radioactivity—conditions that no sane person would choose." Especially in the more high-pressured university and industrial settings, lab directors drive their graduate students and research assistants mercilessly. A result of this is that scientists have created a demanding culture of work: long hours can become a badge of honour, a sign of commitment to research and to the profession.

Inflation in the cost of research has outpaced inflation in the rest of the economy, meaning that the increased money may actually buy less research, if it is possible to measure such things. As a result, most researchers are not buoyant about the state of scientific funding, finding it harder and harder to win competitions for essential grants. In 1945, Vannevar Bush could write *Science: The Endless Frontier*, an optimistic and successful argument for setting up a system of large-scale government funding. In 1991, Leon Lederman wrote *Science: The End of the Frontier*, a pessimistic requiem for that system. The period since World War II has been a boom-time for science, but now that the boom is quieting down, established expectations are being frustrated.

How Much Autonomy Can Science Maintain? The Case of Genetically Modified Potatoes

When, in the summer of 1998, Arpad Pusztai, a researcher at Scotland's Rowett Research Institute, was interviewed on television about his study of genetically modified (GM) potatoes, he hoped that what he had to say would have some impact. He couldn't have guessed how much.

Pusztai's experiment looked at the effects on rats of eating potatoes that had been engineered for insect resistance. He compared rats fed on a diet of GM potatoes with those of rats fed on similar non-engineered potatoes. Rats in the first group had abnormal growth in parts of their intestines and showed the potential for some other worrisome effects. Because, he later claimed, this was potentially a pressing health concern, and because he thought that it would take a year or so for an article on this topic to go through the normal steps of peer review, multiple revisions, and publication, Pusztai decided to make his study public. Besides giving the facts he knew, he spoke bluntly, saying that he himself wasn't going to eat genetically modified crops, and that it was "unfair to use our fellow citizens as guinea pigs."

His words fanned the flames of concern over GM products. European consumers, still living with fears of mad cow disease, were not predisposed in favour of genetic engineering in agriculture, so a study suggesting dangers confirmed many people's suspicions. Consumers wondered loudly where their food was coming from, a number of major supermarket chains and food producers announced that they wouldn't sell genetically modified foods, and the British Medical Association spoke out against GM crops.

Pusztai lost his job. In speaking out, especially before his work had been reviewed by other scientists, he violated one of the central rules of scientific etiquette. Peer review doesn't confer truth on findings, but it is an essential legitimizing stage. Although Pusztai eventually submitted an article, co-written with Stanley Ewen, a pathologist, to the important medical journal *The Lancet*, there was no putting the cork back in the bottle. Before the article was published the Royal Society, Britain's most prestigious scientific organization, did its own review of the research and pronounced it flawed. The U.K. government's scientific adviser criticized Pusztai for not going through the peer review process. And many prominent figures inveighed against him and his work.

The Lancet sent Ewen and Pusztai's article to six reviewers. Four of them said that it should be published on its scientific merits, though they all had criticisms and demands for revisions. One said that it should be published despite flaws in the experiment, to bring the debate into public awareness. And one said that the article should be rejected. The editor then accepted it, on the condition of extensive revisions. Before it appeared, however, the negative reviewer contacted a newspaper and claimed that the article was being published for political reasons; that reviewer thus was violating another rule of scientific etiquette by claiming the sole right to determine the fate of the article.

The experiment was not without its complications. There were only six rats in each group, a small number on which to base large conclusions. Because the GM potatoes were very low in protein the rats were given supplemental protein, though less than an optimal amount. On these and many other grounds Pusztai's study was and is open to questions, criticisms, and alternative interpretations. As we saw in the debates over Mitochondrial Eve it takes only a little prying for scientific studies to become controversial.

Probably more interesting than the technical controversies, though, are the controversies over the behaviour of all concerned. Was Pusztai justified in ignoring science's normal rules about public announcements? Was the Royal Society justified in doing an unsolicited review of an unpublished—indeed unwritten—study? Was the Royal Society biased in its review? Was *The Lancet* publishing Ewen and Pusztai's article on non-scientific grounds? Was the negative reviewer right to make his complaints public? Why did so many bodies become involved and so many people become very personally involved?

This last question has sparked a new level of dispute. Richard Horton, editor of *The Lancet* has suggested an answer that has disturbed many prominent scientists in the U.K. The Royal Society, he claims, has become a lobbyist for science funding. The central strategy in its lobbying efforts has been to stress science's economic value, and so it has made a number of highly visible partnerships with industry. The Royal Society and many of its prominent members have been captured by industrial interests, and support for industry on scientific issues has almost become a reflex. Pusztai's research was pointed in the wrong direction.

As editor of *The Lancet* and author of editorials in the course of the Pusztai affair, Horton is not a disinterested observer. But right or wrong he has expressed a concern felt by many people, both inside and outside of science. As corporate funding increases, as more scientists commercialize their research, and as more industry/university partnerships are being created, science looks less and less disinterested. Scientific voices are losing credibility in the public realm. The autonomy of science may be threatened, then, by the industry that supports it, and by the image that comes from that support.

Purity Is Not an Option, But There Are Better and Worse Compromises

The sciences need some measure of purity if they are to maintain their legitimacy. What makes science valuable is its claim to be a disinterested pursuit of truth. This is not to say that people value all truths for their own sake. Some truths are valuable for solving important puzzles or explaining important phenomena, but more usually, scientific truths are valued for the benefits they might produce.

The strengths of science are undoubtedly intellectual ones. Nobel laureate David Baltimore says that "Science is much better at solving problems of its own devising than those it is asked to

A Utopian representation of ancient science, in the form of pure inquiry in Plato's Academy. Contrast that with Fritz Lang's dystopic image of subterranean infrastructure in his film *Metropolis*. There, science has created a machine that integrates steam, cogs, and workers, all in the service of a ruling class that lives in leisure and luxury above-ground.

solve." Nonetheless, in the course of solving problems of all types scientists have produced any number of practical results. They have contributed to public goods, most prominently medical advances such as vaccines and antibiotics. Science's contribution to military strength has been so important since World War II that in some countries, most notably the United States and the former Soviet Union, the military dominated the funding of science in the 40 or so years following the war. More recently, other industries and other branches of government have stepped up the funding of science for economic reasons: revolutions in the biological sciences, in computer science, and in materials science have been in part fuelled by the promise of economic development and profits.

As we have seen from the sketches in this chapter, when science becomes involved in areas of public interest, it makes compromises—most small, but some large—that may jeopardize its status of purity. Real-world science makes choices about directions in which research will proceed, about questions to be asked, and about images that it will present. Those choices are shaped, usually very subtly, by interests that lie outside science.

Of course, there is no other kind of science than real-world science. Scientists have always had to make choices. These are easier to identify, and sometimes more questionable, when commercial or national interests are at stake. Complete and utter purity, however, was never an option. From the moment science became more than mere curiosity it was shaped by professional interests. From the moment science started to take shape it promised economic benefit, military advantage, and public good.

There have always been better and worse choices. Given its long-term success in enriching our understanding of the world around us, in contributing to technological development, and in maintaining its own autonomy, we can confidently say that it has made more good choices than bad ones. Science's current configuration, which combines high-pressure work environments, larger and more expensive research, new funding constraints, and an increased attention to practical value, is undoubtedly creating enormous strains. So whether science will continue to make more good choices than bad is an open question.

CHAPTER SEVEN

The End of Science?

FRANCISCI
DE VERULAMIO,
Summi Angliæ
CANCELLARII,
Instauratio
magna.

Sin: Pass: sculp.

Multi pertransibunt & augebitur scientia.

Anno

LONDINI
Apud Joannem Billium,
Typographum
Regium.

1620.

•

Is an end to science, or to progress in science, on the horizon? Apparently, many people seem to think so. Their reasons, though, stem from misconceptions of the nature of scientific work. When we apply the approaches of this book to the question, we are apt to see a very long future for science. It is unlikely that there will ever come a time when people find no interesting problems to solve, things to explore, or theories to overturn.

The closing of the twentieth century, a century of tremendous scientific achievements and scientific growth, brought a number of writers to predict the demise of science. The turning of the millennium was apparently for some an ominous marker for all that is modern. Amidst books announcing the end of everything from history to literature were not one but two prominent books in the 1990s entitled *The End of Science*, one of them coming out of the twenty-fifth Nobel Conference. In addition, other prominent books described the collapse of science, an imminent theory of everything, and so on. Such pronouncements have come not only from the pens of today's Oswald Spenglers, expecting a general decline of civilization. Hopes and fears of the end of science have stemmed from mature and sober reflections by practising scientists themselves, and have been captured by writers in close contact with those scientists.

In Chapter One, we examined two powerful myths that have helped to shape the public's understanding of science. On one side is the image of the scientist as a cool clinician, an automa-

◄   The frontispiece for Francis Bacon's *Great Instauration* (1620) shows a ship sailing through the pillars of Hercules. Below the ship is a quotation from the Book of Daniel, "Many shall run to and fro and knowledge shall be increased," a prophecy for the end of the world and the Day of Judgement. Though when Bacon was writing modern science had barely begun, he had a vision of an orderly method by which he thought the project of science would be completed in a matter of years.

ton for investigating nature. On the other is the image of the scientist as a romantic genius driven by intuition and emotional power. In their places we have tried to develop new images of science as a thoroughly human and social activity. These, we argue in this last chapter, point not to the end of science but to a very open future. But first, we must see what prompts the strange millennialism.

Fundamentalist Science: Reaching the End of the Road

To understand how a complex system works investigators usually try to break it down into its constituents. For example, some of the constituents of the developing embryo are the genes that are activated, the cell structures that control transcription of the genes, the proteins that specify which genes are activated, the proteins that are produced, and so on. Developmental biology has been successful in recent years in learning about the protein signals and how they switch genes on and off, how patterns are formed in some simple cases, and much more. Though this is a field where the complexity of the subject matter is daunting, developmental biologists can now claim that they understand many of the key processes.

Scientific understanding often proceeds, as in the case of developmental biology, from the large to the small, from the complex to the simple, and from the surface to the interior. In short, understanding proceeds from what is seen as less fundamental to what is seen as more fundamental. This is reductionism. In its most common usage, "reductionism" has a negative meaning: reductionistic thinking is oversimplified and overzealous in its attempt to see complex phenomena as manifestations of simple and unitary processes. In the nature-nurture debate genetic determinists are called reductionistic by their environmentalist opponents because they believe that everything important about human behaviour can be understood genetically. In contrast to this common usage, we use the term here in a more neutral sense: reductionism assumes that understanding some phenomenon requires exploration of a different level of reality. The new level

Pablo Picasso, *Le Chef d'oeuvre inconnu* (1924).

can be analyzed into subsystems, relationships, or processes, which are then seen to account for and explain the observations at the familiar level. Thus, both sides of the nature-nurture debate are typically reductionistic; whether they are simplistically so depends on the sophistication of their representatives.

Reductionism is often taken to imply that the familiar *is only* that into which it can be analyzed: sexual attraction is just the expression of hormones, which is just the expression of genetically coded imperatives, and so on. As such, reductionism appears to create a hierarchy among objects—and a parallel hierarchy among the sciences that study them. In the hierarchy of sciences the point on which all else rests presumably is something like the most basic particle physics, which investigates the nature and behaviour of the constituents of matter. Victor Weisskopf, a former director of CERN (the European Organization for Nuclear Research), gave a convenient term to this view: *fundamentalism.* For Weisskopf, "all sciences are at the end a branch of physics," because everything in the world can be reduced to

A model of the structure of DNA.

quantum mechanics. But one doesn't have to look to physicists to find statements of the hierarchy. When James Watson and Francis Crick, co-discoverers of the structure of DNA, called it "the secret of life" or when geneticists say that the Human Genome Project will result in "a complete guide on how to build and run the human body," they imply that DNA sits in a more important place in the hierarchy than other biological structures.

The former director of the Fermi Accelerator Laboratory, Leon Lederman, once mused that the aim of science, finally within our sight, is to reduce all laws governing natural phenomena to one equation that will fit on a T-shirt. Lederman, then, is not only a fundamentalist, but an optimistic one, predicting the end of science because physics is closing in on the last equation. The tower will soon reach unto heaven. After that, everything else is just mopping up. He is not alone among prominent physicists in being optimistic. Stephen Hawking, the wheelchair-bound Lucasian Professor at Cambridge, famous for his *A Brief History of Time*, thought as recently as 1974 that there was chance that physics would pack up by the end of the twentieth century, having solved the important problems. Nobel laureate Steven Weinberg is more appreciative of the immense scope of the sciences, but he titled a recent popular book *Dreams of a Final Theory*.

Optimistic fundamentalism can lead easily to pessimism. A failure by governments to support the next-generation accelerator or the next-generation telescope may mean an end to fundamentalists' dream of elaborating the *final theory*, at least in their lifetimes. Weinberg writes:

> It is a pity that new accelerators and telescopes happen to be expensive, but not to build them would mean that science must renounce the highest of its objectives, the discovery of the laws of nature. One sometimes encounters the puritanical opinion that a failure to build the next generation of accelerators will be compensated by an increased cleverness in the use of theory, string, and sealing wax. But the decades of stagnation in the study of gravity show what must happen to even the most exciting subject without the pressure of data.

Science, then, is coming to an end because basic physics is not getting the funding it needs.

Fundamentalism, we can see, is a key source for thoughts about the end of science. Fundamentalists believe that what is most important in science will come to an end because the final discoveries are just around the corner, or will come to an end if they are not given the resources to look around the corner. Though to many it may seem like parochialism, there is a noble dream in fundamentalism. It is, however, a wrong-headed one. At its core it misinterprets the successes of analytical reductionism and the power of spare and abstract mathematical representations for a too-simple and too-inevitable order. Our next task is to examine fundamentalism, and in so doing sever the connection between reductionism and millennialism regarding science.

Generously endowed by the Carnegie Institution, in 1918 the Mount Wilson Observatory acquired the largest telescope in the world, with a 100-inch mirror. From then on astrophysics became a large-scale science.

A Hole in the Desert: The Superconducting Supercollider

The Superconducting Supercollider (SSC) was to be the ultimate experimental apparatus, capable of reproducing for a brief moment some of the conditions existing during the "Big Bang." Approved by President Reagan in 1987, this mammoth accelerator was to be the largest scientific project in history, operating along a 53-mile racetrack-shaped tunnel at Waxahachie, Texas. It would cost about $8 billion and take 10 years to build. Fitting for such a grand project, hyperbole was not absent from the vocabulary of its promoters. "If we can be suc-

An historic meeting at the White House. On January 29, 1987, Alvin Trivelpiece, director of the U.S. Energy Department, briefed President Reagan on the need to allocate $8 billion to the SSC project. The President was apparently convinced, and answered paraphrasing Jack London: "I would rather be a superb meteor, every atom of me in magnificent glow, than a sleepy and permanent planet."

cessful with the SSC," declared Robert Roe, the New Jersey Democrat who headed the House Committee on Science, Space and Technology, "we can forever revolutionize knowledge in the world."

Eight billion dollars was so much money that it drove a wedge among scientists, and among physicists in particular. In an effort to win the favour of other scientific communities, Steven Weinberg introduced a colourful, though aggressive, metaphor:

> Instead of feuding with one another for public favor, it would be fitting for scientists to think of themselves as members of an expedition sent to explore an unfamiliar

but civilized commonwealth whose laws and customs are dimly understood. However exciting and profitable it may be to establish themselves in the rich coastal cities of biochemistry and solid state physics, it would be tragic to cut off support to the parties already working their way up-river, past the portages of particle physics and cosmology, toward the mysterious inland capital where the laws are made.

Physics needs big instruments because without them it cannot discover the basic laws. Everything else is little more than tourism, but fundamental physics is on an expedition of colonial discovery.

Of course, Weinberg might have *intended* to end feuding over funds, by belittling all other fields he only reinforced the feuds. James Krumhansl, president of the American Physical Society, took particular issue with what he saw as the arrogance of particle physicists, when he contrasted the ease with which the Hubble Space Telescope received funding and the difficulties faced by the SSC. About the former he noted that physicists in "other fields never belittled the value of the Space Telescope or argued that the funds should be diverted their way." The difference, he says, lies in "respect in the community for the way the cosmologists went about securing support for their program. They did not represent discoveries from other subfields of physics as their own by inference."

Another of the opponents was Freeman Dyson, Professor of Physics at Princeton University and a well-known essayist. "The advocates of the SSC often talk as if the universe were one-dimensional, with energy as the only dimension," he wrote. According to Dyson, over the history of particle physics the most important experimental advances have been split between achievements of greater energy, achievements of greater accuracy, and achievements of greater rarity. "There is no illusion more dangerous," he concluded, "than the belief that the progress of science is predictable. If you look for nature's secret in only one direction, you are likely to miss the most important secrets, those which you did not have enough imagination to predict."

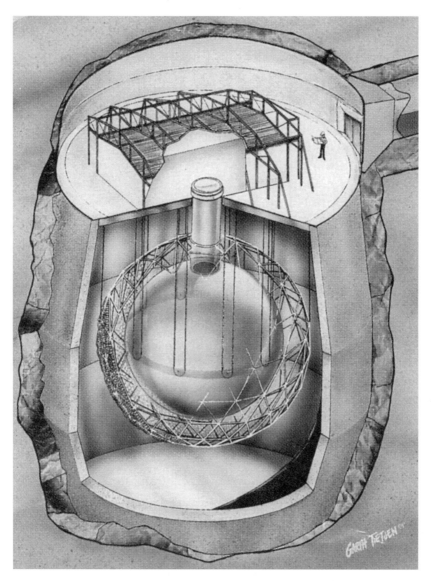

The Sudbury Neutrino Observatory (SNO), deep underground in an old mine in Sudbury, Canada, gathers data on a most elusive elementary particle, the neutrino. Two thousand meters of solid granite above the observatory shields out other cosmic rays, but leaves neutrinos largely unaffected, so that a minute percentage of them can leave their marks in a waiting pool of heavy water. Data coming from SNO solved in 2001 the 30-year-old puzzle of why some neutrinos were missing from the sun; it also provided important information about the sun and the future evolution of the universe.

Many of the SSC's political supporters had coveted it for vastly different reasons. For some it was judged in terms of national prestige, for others in terms of job creation or purely in terms of local election politics. Somewhere, "the quark got confused with the pork," a columnist wrote. For example, one week before the crucial vote, Robert Roe, the same New Jersey representative who had promoted the SSC as a means of protecting U.S. scientific pre-eminence, threatened to oppose the project unless more money was restored to his own state's scientific project: the magnetic fusion lab at Princeton. Roe's resistance was averted by a last-minute move by Jim Chapman, a Democrat representative from Texas who engineered a transfer of funding from energy and water programs to the Princeton laboratory project.

The project succumbed to political pressure and budget considerations that had little to do with scientific arguments. In the context of deepening recession in the early 1990s and annual budget crises in Washington, the SSC's billions started to look like nothing but pork, especially since so much of that money would be spent in Texas. Money spent on the SSC looked like money *not* spent on health and education or on research projects elsewhere. Since World War II, physicists in the United States had grown used to full funding for their large projects, but the SSC was too large and wasn't clearly enough connected with national goals, especially with the ending of the Cold War. Today, perhaps tragically, all that is left of the SSC is a very long hole where it would have lain.

The Incompleteness of Reductions

When scientists look for a final theory, or even attempt to explain a less fundamental domain in terms of a more fundamental one, they do not really expect to *replace* the less fundamental by the more fundamental. Nature is too complex for that. Although they draw on it in their work, chemists cannot use quantum mechanics as a calculating device to solve most interesting chemical problems, because even with the fastest of computers the mathematical calculations are too many and too large. Instead chemists need to use concepts formulated roughly at the level of the phenomena that interest them. The same applies to geologists, geophysicists, molecular biologists, ecologists, sociologists, and others: while theories and results at lower levels can be useful, they rarely can be used to do the work of theories and results at higher levels.

Reductions aim at something weaker than replacement. They might aim at complete explanations, but successful reductions do not typically provide full accounts of the things being reduced; instead they elucidate key relationships. Michael Polanyi, a chemist and solid-state physicist, described this when he argued that physics and chemistry cannot explain living organisms: "Lower levels do not lack a bearing on higher levels; they define the conditions of their success and account for their failures, but they cannot account for their success, for they cannot even define it." In order to account for the success of a higher level in terms of a lower level, one has to be able to reconstruct the higher in terms of the lower. That is a rare achievement, because it demands that everything people think is important about the higher level be predicted by what is known about the lower. Solid-state physicist and Nobel laureate Philip Anderson thinks that fundamentalism suffers from looking in only one direction:

> The ability to reduce everything to simple fundamental laws does not imply the ability to start from those laws and reconstruct the universe. In fact, the more the elementary particle physicists tell us about the nature of the fundamental laws, the less relevance they seem to have to the very real problems of the rest of science, much less to those of society.

Let us return briefly to the developing embryo. Lewis Wolpert is a prominent developmental biologist and a writer on science. Despite the fact that he thinks we now understand the basics of development, he is still cautious:

> Given a total description of the fertilized egg—the total DNA sequence and the location of all proteins and RNA—could one predict how the embryo will develop?

This is a formidable task, for it implies that in computing the embryo, it may be necessary to compute the behavior of all the constituent cells. It may, however, be feasible if a level of complexity of cell behavior can be chosen that is adequate to account for development but that does not require each cell's detailed behavior to be taken into account.

We should note the careful language Wolpert chooses. He is not saying that biologists will ever be able to descriptively *reproduce* the developing embryo, but rather that they will eventually find a style of description within which important features of development can be explained. Wolpert goes on: "We will, however, understand much more than we can predict. For example, if a mutation were introduced that altered the structure of a single protein, it is unlikely that it will be possible to predict its consequences." Success in developmental biology does not mean being able to compute the embryo in a detailed way, and it does not mean being able to predict the effects of changes. Instead, it means understanding the central causes of normal processes. *Understanding*, we should note, is something that people do; it demands good representations, but not a perfect reproduction of nature.

Multiple Autonomies

The fundamentalist picture of all science depending on "basic" physics is misleading in a different way. Not even all *physics* depends on basic physics. To continue our procession of Nobel laureates, Ilya Prigogine argues that though both are extraordinarily well confirmed, thermodynamics *contradicts* particle physics. Thermodynamics demands the asymmetry of time, seen in the irreversibility of scrambling an egg, yet at the "fundamental" level all processes are reversible and symmetrical in time. Although there are standard strategies to deny that this is a real contradiction, Prigogine claims that those strategies amount to little more than denial. We could take from his argument that one or both of thermodynamics and particle physics need to be changed. Or we could accept that both are approximately right, but that the concepts they employ are so different, and the processes they describe so distinct, that both can be approximately right while being in conflict. Asymmetry at one level coexists with symmetry at another. If so, then the one cannot be reduced to the other.

This lesson does not have to stem from apparent contradictions within physics. For other sciences the reductionist's dream is so far from being possible that contradictions cannot even be seen. The behaviours of large and complex systems, it turns out, are not to be understood in terms of simple extrapolation of the properties of a few particles. Instead, at each level of complexity entirely new properties appear, and the understanding of the new behaviours requires research as fundamental in its nature as any other. For each domain entirely new laws, concepts, and generalizations are necessary, requiring inspiration and creativity to just as great a degree as in the next domain. And experimentalists, as we saw in Chapter Five, can even create new phenomena to study, phenomena that might need new laws.

When we look to climatology, geology, medicine, psychology, and biology, we see sciences that study complex systems, and do so largely within their own terms. Darwin's theory of evolution by natural selection is a nice example, because it is so simple, familiar, and clearly independent of basic physics. It is an interesting and provocative theory in and of itself, employing concepts, such as fitness, adaptation, reproduction, that have little connection to physics. If the laws of physics were quite different from what they are, the theory of evolution by natural selection might remain exactly the same, with no loss of accuracy or interest. It is, for example, possible to simulate evolution on a computer, by letting strings of computer code mutate and replicate. Unsurprising to evolutionists, the theory of evolution applies not only to flesh-and-blood organisms, but to electronic ones. Of course, the laws of physics also apply to both living organisms and to electronic ones, but they do so in radically different ways. Thus we can see that physical laws play no part in the reasons why the theory of evolution applies to widely divergent types of organisms: the theory of evolution applies to structures with certain forms, which might have completely different compositions.

From the point of view of the evolutionary biologist, then, the laws of physics are largely irrelevant. Reductionism is flawed if it claims that all the laws of biology must depend on those of physics. But if the theory of evolution does not rest upon theories of molecular biology, of chemistry, of physics, then there is a sense in which it describes an *autonomous* level of phenomena, not part of the reductionist's hierarchy: the theory of evolution is itself a fundamental theory. This is not to say that reductionistic strategies are useless in evolutionary biology. Learning about processes of reproduction at the level of DNA, for example, has enriched the field immensely, introducing new phenomena to be studied, showing constraints in processes of evolution, and so on. However, these strategies do not show that evolutionary biology is just physics, and they don't support the metaphysical reductionism that foretells the end of science. Rather they show one can learn much about something by looking at its parts.

It is, then, not possible to reduce all phenomena to a single basic level, because there are many autonomous levels in nature. Autonomous objects and processes at higher levels have constituents, but the nature of the constituents is less important than how they are organized. Autonomous objects and processes have their own fundamentality, in whatever fields they are found. And thus the hierarchy that fundamentalism assumes proves multiple, ending not at a single point but at many. There are different sciences, and not just branches of a single science.

● ●

Not Just Dualism

The flourishing field of neurobiology has developed fascinating and rich descriptions of the workings of individual neurons, the connections between them, and large-scale patterns of brain activity. As a result we know considerably more about the workings of nervous systems and brains. PET scanners can show localized brain activity associated with particular types of

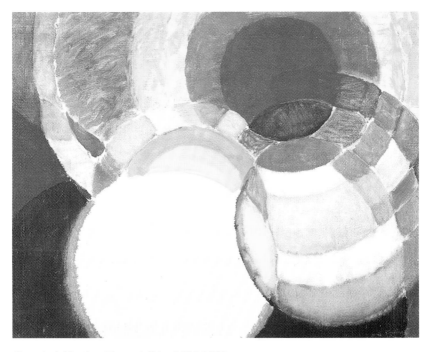

Frantisek Kupka. *Newton's Discs* 1911-1912.

One of the most powerful demonstrations of reductionist thinking at the colour level is given by Frantisek Kupka's work. The Czech artist was very aware of the inspirational power of science and frequently referred to Ogden Rood's *Modern Chromatics* work, which he preferred to Newton's theories.

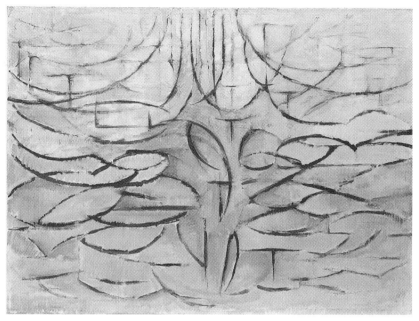

In a series of paintings of the same subject—a single tree—over a period of four years, Mondrian demonstrated his evolution towards reductionist abstraction. In *The Red Tree*, the first of the series, the painter is deeply committed to the Fauve landscape inspired by Vincent Van Gogh, in which the impact of colour is emphasized. In the successive treatments, colour disappears, and Mondrian gradually removes more and more evocative descriptive elements.

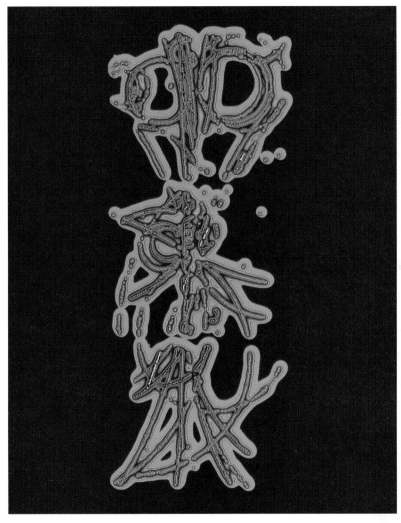

Asked to comment on a possible scientific inspiration to his own work, the painter Hartung said: "I have a horror of painters who try to depict astronomical or physical facts ... if these things penetrate into your spirit, if they take part in the formation of your thoughts, well and good but if anyone sets himself to paint the microbes he sees down the microscope—he would do better to paint the pretty girls in Montparnasse or Montmartre." Nonetheless, there is a strong affinity between Hartung's *Peinture* (1963) (on the next page) and tracks of particles in bubble chambers.

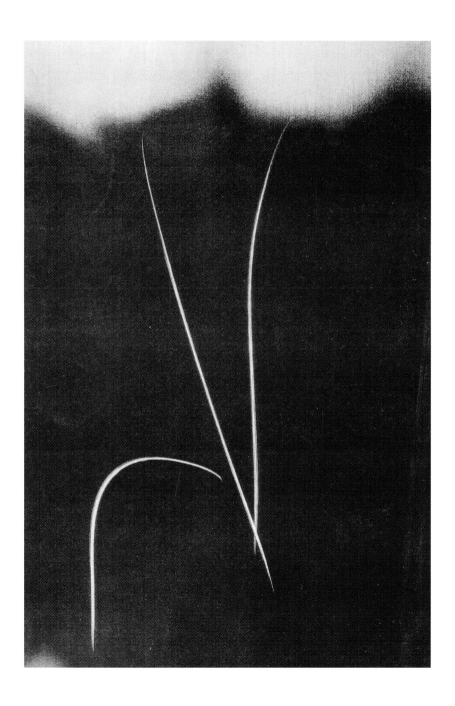

Vassily Kandinsky: *Improvisation V* (1911). Analysing Kandinsky's work, Pierre Volboudt, one of his biographers, wrote, "Almost before his eyes, the image of the world changed. The splitting of the atom introduced a new element of uncertainty. 'Reality' became unstable for Kandinsky, vacillating between appearances, which were illusory, and mental certainties that were not reflected in the visible world. Matter was crumbling, disintegrating into particles of energy, only to re-emerge as fields of force, equilibriums suspended in the 'vast open void.' It was this 'reality' that was to be reflected in his paintings, forms floating unsupported in space. For Kandinsky, Niels Bohr was the prophet of an apocalypse of the visible world. 'I could,' we read in the painter's notes, 'have seen a stone dissolve in the winds and evaporate, and I would not have been surprised.'"

Does this mean that scientists are artists and *vice versa?* No. For every similarity between science and art there are many differences in techniques, skills, audiences, and rigour. But from some angles we can see that the intellectual strategies for making such revelations and revolutions are not wholly different in these different realms.

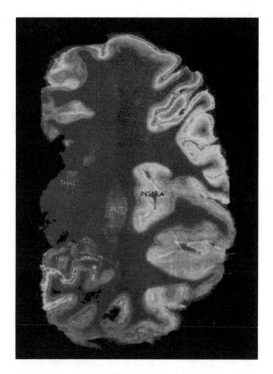

PET scans can show blood flows in the brain, which
enables researchers to see brains in action.

stimuli: computer images like the one here show different parts
of the brain used for different activities. But neurobiology has
neither elbowed traditional psychology out of the picture, nor
even explained key psychological results. Instead, psychologists
continue to seek to understand human behaviour in terms of
beliefs, fears, group dynamics, and so on; in fact, there are more
psychological questions all the time, some of them even
spawned by neurobiology. Reductionist strategies are successful
in the brain sciences, but they do not reduce away minds.

Most people become very uncomfortable with fundamentalism when it gets close to the human mind, and they are probably
justified. Surely the mysteries of thought will never be explained
in purely physical terms! But if the arguments of this chapter are
correct, then minds are not special in being resistant to reduction.

Psychological phenomena have a certain autonomy and logic, which requires that we analyze them in terms of meaning, and not just neural activity. Neurobiology neither replaces nor even fully explains other understandings of individual and social behaviour. But most other phenomena also have some degree of autonomy: analysis into components is immensely valuable, but it rarely dispenses with the phenomenon or its properties. Thus, we should not be content with a simple dualism that divides the world into mind and matter, but instead find comfort in a robust pluralism in which minds are objects particularly dear to us.

●●

Final Theories and the Art of Science

So far we have been talking about fundamentality. But what about finality? Is there reason to believe in truly final theories? Here the dominant myths of science obstruct the view, because they don't make space for the real creativity and work of making laws. The myths of the genius and the computer don't make room for strategies for starting scientific revolutions, the possibilities of reopening closed questions, or even the skills in creating new systems for study.

If there were laws that appeared simply to sit by themselves at the bases of things, would they not become targets for would-be revolutionaries, some of them espousing fundamentalism, keen to find some vantage point from which to create yet more basic theories and laws? Can we be sure that there are laws that cannot be opened up, turned inside out, seen from another view? Is it not possible that under each foundation scientists could build another if they tried hard enough? Or that they could continually refine patterns of thinking, creating new problems to solve?

Among other things, the fundamentalist assumes that theories and laws are fully things of the natural world, that they are structures inhering in the natural world. Theories and laws are not imposed on the world, but, even when they are true, they are human descriptions of how things that interest us behave. To see those descriptions *in* the world is to make a category mistake, to mistake the representation for what it represents. That is not to say that theories and laws are simply made up or that they are false. On the contrary, a true theory or law describes some domain, as a good representational painting depicts something in the world. There is every reason to believe that the majority of accepted scientific knowledge is true, that it captures order in nature. However, if there is no obvious limit to creativity in painting—or poetry or sculpture or banking or any other creative human endeavour— why should we assume that there is a limit to creativity in science, even fundamental science? New perspectives, new anomalies, and new techniques of representation can keep fuelling the activity of theory-building. Truth is not a single stopping point.

The fundamentalist forgets that science is a human activity. Concepts, theories, models, and laws are constructed because they enable people to understand particular domains. A truth about one domain does not have in it all of the details of all of the other connected domains. An excellent photograph when enlarged becomes grainy. Those grains are little many-coloured lies, artifacts of the limitations of the equipment. On the one hand, those limitations make for a more versatile, more useful camera. And, on the other, it is not at all obvious that a camera without those limitations would always, or even ever, take better pictures—perhaps this is another lesson we could take from the history of art in the twentieth century.

Pessimism and the Cultural Position of Science

No scientific theory, however basic, could replace all others. There is no necessary hierarchy in the sciences, because there are autonomous levels in nature. And final theories are not to be expected. These observations remove some misplaced hopes for the completion of the scientific project, but they do not remove all fears of the end of science. They respond to the optimistic fundamentalist, but not to the pessimist.

Even many scientists who would deny that physics is fundamental in any interesting sense can become pessimistic. Most fields felt some slowing of funding for basic research in the 1980s and 1990s. That is, the problem is not just one of the building of bigger telescopes and accelerators: science, after having grown exponentially through its history, has reached a point of crisis. As a result many scientists complain of a general decline in civilization, evidenced through its lack of support for creative and productive endeavours.

Of course, human creativity *may* be limited when it comes to building foundations. It might reach a point where it takes too much effort to question the solidity of an existing edifice, or too much effort to dig any deeper. Without new big tools to produce striking new data, stagnation might set in. Just as there is no intrinsic reason why scientists shouldn't be able to keep on discovering novelty, there is no intrinsic reason why scientists *should* be able to. But to make that case the pessimist has at least as much work to do as does the optimist.

If funding dried up, the sciences would become unrecognizable. Lost would be the current social structures, which depend on journals, conferences, and funds to support post-doctoral researchers, technicians, graduate students, and others. Lost would be experimentation with more than sealing wax and string. But funding, while growing more slowly than it once was, and changing form, is not disappearing. In and of itself the changes may have profound effects, and some of them may be deplorable, but there is no reason to see in them the end of science. Increased funding for commercial science will undoubtedly funnel much research into the most profitable channels. The best graduate students may choose fundable disciplines and problems over unfundable ones, but that would not be a new phenomenon—fundable problems nevertheless allow researchers to use their skills and intellect. One biochemist interviewed on this topic said that he left the university environment to join a biotechnology company because the stability of funding allowed him to focus on hard problems, problems that should be considered basic science: "The most astute questions that were asked me by any of the people at any of the seminars I had given were, in fact, from the people at [the company] who ... really seemed to be in a position to have the resources available to make the most use of their talents, unlike the struggle in the

The construction of large scientific instruments has often tested the limits of contemporary technology. Here the dimensions of the underground accelerator rings at CERN outside Geneva seem to dwarf the runway of the international airport.

academy." So while we should question some of the priorities of commercially funded science, it is clearly science.

Probably the most major shift in the funding of science has come because of a confluence of some unrelated factors. The ending of the Cold War dramatically reduced interest in the West and East for supporting the physical sciences. One of the lessons of World War II was that even basic physics and chemistry could lead to military applications: radar and the atomic bomb were the two most notable examples, but there were many others. So whenever they could even tenuously and remotely link projects to defence interests, physical scientists were likely to receive support. When the Soviet Union collapsed in 1989, the perceived imperative for that support dropped away. Meanwhile, molecular biology was changing from an esoteric science to one that had implications for all of the life sciences, and for medicine in particular. Scientists went from painstakingly eking out discoveries in this field to confidently planning major initiatives, of which the Human Genome Project is the largest. The results of this have been remarkable not only in molecular biology, but in many of the fields that border it. Not least of these results has been the attraction of support, from governments and other agencies, and also from drug companies and companies that have entered the new field of biotechnology. So the life sciences gained what the physical sciences lost, both in terms of money and in terms of perceived importance for the future.

Science, then, is not a static entity. Researchers work to develop new theories, to design new tools, and to find new niches. Even when it becomes prohibitively expensive or prohibitively difficult to pursue one or another avenue of research, other avenues remain.

There is one last important source for speculations about the end of science: a change in perceptions of science, away from myths like the myth of the computer and the myth of the genius. The topic of the 1989 Nobel Conference shows some of the fears produced in making science "human." "Science as a unified, universal, objective endeavour is currently being questioned.... What, given this development, should science mean to society? It can no longer mean, as it meant to modernism, the stable guide in a world of humanistic instability. Is this new understanding forecasting The End of Science?" Does the end of faith in science mean the end of science?

In challenging an aging ideology of science are we helping to end not only the ideology but the pursuit of science itself? Our very answers to the optimistic and pessimistic fundamentalists may look as though they undercut the cultural authority of science. If this were true, then we would have failed in one of our purposes: at the beginning of this book, we wrote that we hoped that we would make science look a little less miraculous, but no less impressive.

The myths of the computer and the genius make science an otherworldly activity, by making scientists appear inhuman. Because they are so clearly myths, they lend little legitimacy to science. In contrast, we have emphasized the humanity of science, its continuity and comparability with other activities. Thus we have tried to portray science not as an object of veneration from afar, but as an art for appreciation from next door:

- Science is an art because it produces representations. Representations are not part of nature, but descriptive of it. Scientific theories are not intuited directly, but are developed. Thus, during revolutions there are large changes in perspective, changes in styles of representation and interaction.

- Science is an art because scientific reasoning is not computation. Reasoning is a social product. Scientific communities learn how to think in useful and structured ways about their subject matters.

- Science is an art because experimentation is hard work that involves considerable skill, not simply sitting back and observing what happens. Experiments are interactions of people and nature. The knowledge that results from experiments is not merely knowledge of the surfaces of nature, but knowledge of structures of possibility.

- Science is an art because it takes place in the real world. The work of research is inextricable from the work of convincing agencies to fund the research and the work of convincing others to pay attention to ideas.

None of this provides any reason to doubt the knowledge produced by the sciences. Instead, it provides a better understanding of the solidity of scientific knowledge, which is based in the efforts of many brilliant and hard-working people. Revoking public myths regarding science should allow science to gain, not lose, authority. While the end of public faith might mean the end of science, the end of an ideology only means the end of blind faith, which is always a too fragile basis on which to build foundations.

Conclusion

For many scientists, the successes of reductionism incline them to fundamentalism. Fundamentalists assume that the most exciting and valuable research is the discovery of something that lies hidden, and perhaps will always lie hidden at new depths. They imagine one big discovery that will explain everything. They see only a tower being built towards heaven. If the tower can be made to reach heaven, then science is over, a success. If it cannot, then science is a failure, another tower of Babel. That sort of science is essentially a religious activity, an attempt to find the equation that can stand in for God. But that equation, if it could ever be found, would explain almost nothing of the world. In placing the only important goal of science so far away the fundamentalist may tend to ignore the excitement and value in creating richer pictures of, and finding new problems in, the world closer to hand. That sort of research shows no sign of end, for surprising discoveries are made all the time. The future looks most promising, then, for those who subscribe to a pluralistic vision, seeing real problems everywhere they look. And if we recognize science as a human activity the future is more promising yet.

Francis Bacon, writing in the 1620s, thought that if European researchers followed his methods and his ideas, scientific knowledge would be complete within a few generations. Some 15 years later, René Descartes wrote almost exactly the same thing about his methods and ideas. Although those claims were quickly made to look ridiculous, similar grand pronouncements have been made by almost every generation since then. In the late nineteenth century, for example, they became commonplace. In 1888 the astronomer Simon Newcombe said that "It would be too much to say with confidence that the age of great discoveries in any branch of science has passed by; yet so far as astronomy is concerned, it must be confessed that we do appear to be fast approaching the limits of our knowledge." In 1894 the physicist Albert Michelson, whose results helped pave the way for the theory of general relativity, said that "our future discoveries must be looked for in the sixth place of decimals." And in 1900 Lord Kelvin said that "There is nothing new to be

Pieter Breugel, *Tower of Babel* (detail).

discovered in physics now. All that remains is more and more precise measurement."

We might speculate about the conditions that prompt people to make such claims and the conditions that allow the claims to achieve prominence. Does the end of science look closer in the wake of a major revolution? Are claims of the end of science really calls to arms, calls for scientists to rally around some project or some vision of the nature of science? If so, might we expect to see claims of the end of science precisely when there is disunity of purpose or approach among scientists? However we answer such questions, it should be clear that no prediction of the end of science has yet been borne out.

Indeed, perhaps science has only just begun.

In writing this book we drew on a number of articles and books, though more than that we drew on our experience teaching about the practice of science. For readers interested in looking further on the topics we have covered, or interested in the works from which we have drawn, we have compiled this brief bibliographic essay. Besides our own sources, we have included only a small sample of works, concentrating on ones that are particularly important.

There are a number of other books that have a similar scope as does this one. Jacob Bronowski's *Science and Human Values* (New York: Harper & Row, 1965) is a classic, but does not have the advantage of the past three decades of thinking about science. It does not even reflect the insights of Thomas Kuhn's groundbreaking work *The Structure of Scientific Revolutions* (Chicago: University of Chicago Press, 1962), let alone more recent important works. Bruno Latour's *Science in Action: How to Follow Scientists and Engineers Through Society* (Cambridge, MA: Harvard University Press, 1987) is a book of large theoretical scope and considerable interest. It is occasionally difficult to read, but Latour's Gallic prose makes it always charming. *The Golem: What Everybody Should Know About Science*, by Harry Collins and Trevor Pinch (Cambridge: Cambridge University Press, 1993) is an irreverent treatment of controversies and the contingency of scientific knowledge. Readers interested in a sampling of recent scholarship on science should see *The Science Studies Reader*, edited by Mario Biagioli (New York: Routledge, 1999).

Chapter 1:
The Computer and the Genius

There are many introductory texts on the history of science. A recent general overview in a style comparable to the book you are holding is Peter Whitfield's *Landmarks in Western Science: From Prehistory to the Atomic Age* (New York: Routledge, 1999). A book that emphasizes science's contribution to modernity is John Marks's *Science and the Making of the Modern World* (London: Heinemann, 1983). For the seventeenth century Steven Shapin's *The Scientific Revolution* (Chicago: University of Chicago Press, 1996) and Peter Dear's *Revolutionizing the Sciences* (Princeton: Princeton University Press, 2001) are short and readable discussions of general issues and of historiographical issues. *Wonders and the Order of Nature, 1150-1750* (New York: Zone Books, 1998), by Lorraine Daston and Katharine Park, is simultaneously beautiful and a scholarly exploration of the place of the wonderful and exotic, and their contribution to science, in early modern Europe.

Lee Caplin's *The Business of Art* (Englewood Cliffs, NJ: Prentice-Hall, 1989) has a number of interesting essays on, as its title advertises, the business of art.

There are very few accessible treatments of the place of science in popular films. For an overview of science fiction films, which are often only peripherally concerned with science, see John Brosnan, *The Primal Screen: A History of Science Fiction Film* (London: Orbit, 1994). Per Schelde's *Androids, Humanoids, and Other Science Fiction Monsters: Science and Soul in Science Fiction Films* (New York: New York University Press, 1993) explicitly discusses the scientist in science fiction movies. Schelde thinks that the dominant view is of the scientist as romantic genius, which serves as a counter to the supposedly dominant view of science as staid and orderly—science fiction directors are rebellious.

Chapter 2:
Painting the Invisible

The story of the rise of quantum physics has been told many times, including by participants and their students. George Gamow's *Thirty Years that Shook Physics: The Story of Quantum Theory* (Garden City, NY: Doubleday & Company, 1966) is largely arranged as a series of biographical sketches of people that this physicist knew or knew of. Emilio Segrè, a Nobel laureate in physics, is more thorough than Gamow, but similarly focuses on individuals in *From X-Rays to Quarks: Modern Physicists and Their Discoveries* (New York: W.H. Freeman and Company, 1980). C.P. Snow's *The Physicists* (New York: Little & Brown, 1981) is the source of more of our information. As the title indicates, Daniel J. Kevles focuses on America in *The Physicists: The History of a Scientific Community in Modern America*, 2nd ed. (Cambridge, Mass.: Harvard University Press, 1995). Nonetheless, his early chapters give some indication of the forces shaping the development of a community of physicists in the twentieth century.

Paul Forman's 1972 scholarly article, "Weimar Culture, Causality, and Quantum Theory, 1918-1927: Adaptation by German Physicists and Mathematicians to a Hostile Intellectual Environment," *Historical Studies in the Physical Sciences* 3: 79-108, is the source of our comments about Oswald Spengler and the mood in Weimar Germany. That article has been extremely controversial, though in our opinion there has never been a successful challenge of Forman's key pieces of evidence or of his argument. Mara Beller's *Quantum Dialogue: The Making of a Revolution* (Chicago: University of Chicago Press, 1999) qualifies Forman's point considerably; it is also a fascinating and detailed study of the way that dialogues shaped quantum mechanics and its reception.

The article "Do Angels Have Bodies? Two Stories about Subjectivity in Science: The Cases of William X and Mister H," by Hélène Mialet in *Social Studies of Science* 29 (1999: 551-82), provides a very different analysis of genius from ours. Mialet argues that scientific genius is a matter of becoming the right sort of node in a network, and of having a "disembodied" relation to one's material.

Linda Henderson has written an interesting book on early modern art as representing a "fourth dimension," a popular topic that became relevant in connection with Albert Einstein's theory of relativity: *The Fourth Dimension and Non-Euclidean Geometry in Modern Art* (Princeton, NJ: Princeton University Press, 1983). Also valuable is P.C. Vitz and A.B. Glimcher's edited volume *Modern Art and Modern Science: The Parallel Analysis of Vision* (New York: Praeger, 1984). The quote from Gerald Needham on stereoscopes comes from his contribution to that volume.

Lionello Venturi's comments are in his *Cézanne* (Paris: Skiva, 1985). The story of the photography of motion is in Marta Braun, *Picturing Time* (Chicago: University of Chicago Press, 1992). The quotations from Duchamp and Mondrian are in John Russell, *The Meaning of Modern Art* (New York: HarperCollins, 1991). Further information about Mondrian came from Serge Lemoine, *Mondrian* (New York: Universe Books, 1987).

Chapter 3:
Logic and the Construction of Reason

Harry Collins has done much thinking about "tacit knowledge" in the sciences. The locus classicus (already) is *Changing Order: Replication and Induction in Scientific Practice*, 2nd ed. (Chicago: University of Chicago Press, 1991). For another source, one from within the natural sciences, see Michael Polanyi's *Personal Knowledge: Towards a Post-Critical Philosophy* (London: Routledge & Kegan Paul, 1958).

A general discussion of non-Darwinian theories of evolution can be found in Peter Bowler, *The Eclipse of Darwinism: Anti-Darwinian Evolution Theories in the Decades around 1900* (Baltimore: Johns Hopkins University Press, 1983). To appreciate Darwinism we recommend going straight to the source, Charles Darwin's *On the Origin of Species* (originally published in 1859, and available in a number of modern and facsimile editions). Darwin's *Origin* is a wonderfully insightful work, which many biologists still find useful and interesting to read. Darwin's later work on the evolution of humans and other problems, *The Descent of Man: And Selection in Relation to Sex* (1885), is more difficult and less successful.

Fleeming Jenkin's review of *On the Origin of Species* was published in *The North British Review* 46 (1867): 277-318. It is republished in David Hull, *Darwin and His Critics: The Reception of Darwin's Theory of Evolution by the Scientific Community* (Cambridge, Mass.: Harvard University Press, 1973), 303-44. Kelvin's own contribution to the age of the earth debate is reprinted in his *Popular Lectures and Addresses* (London: Macmillan, 1894), 10-72.

Steven Jay Gould is a biologist who stands outside the mainstream Darwinian consensus. His *Ontogeny and Phylogeny* (Cambridge, Mass.: Harvard University Press, 1977) is a study of the out-of-fashion hypothesis that the development of organisms is intimately connected with their evolutionary history. The result is a worthwhile exploration of some aspects of development, paleontology, evolutionary biology, and their histories. George Gaylord Simpson's important synthetic work on paleontology is *Tempo and Mode in Evolution* (New York: Columbia University Press, 1944). D'Arcy Wentworth Thompson's treatise on patterns in development and morphology is *On Growth and Form*

(Cambridge: Cambridge University Press, 1917). There are several editions still in print, including an abridged edition and an inexpensive Dover edition.

Robert Jensen provided the argument for our vignette on the making of modernism in his *Marketing Modernism in Fin-de-Siècle Europe* (Princeton, NJ: Princeton University Press, 1994).

Chapter 4:
Controversies and Consensus

One of the best and most accessible books dealing with scientific controversies is *The Golem: What Everyone Should Know about Science* (Cambridge: Cambridge University Press, 1993), by Harry Collins and Trevor Pinch. There are many articles and other books worth reading on this topic, including the collection edited by Dorothy Nelkin, *Controversy: Politics of Technical Decisions*, 3rd ed. (Newbury Park, Calif.: Sage Publications, 1992). A very enjoyable book on debates about human evolution is Roger Lewin's *Bones of Contention: Controversies in the Search for Human Origins* (New York: Touchstone, 1987).

The key articles on Mitochondrial Eve and molecular anthropology to which we refer are: Rebecca L. Cann, Mark Stoneking, and Allan C. Wilson, "Mitochondrial DNA and Human Evolution," *Nature* 325 (1987): 31-36; Linda Vigilant, Mark Stoneking, Henry Harpending, Kristen Hawkes, and Allan Wilson, "African Populations and the Evolution of Human Mitochondrial DNA," *Science* 253 (1991): 1503-07; David R. Maddison, "African Origin of Human Mitochondrial DNA Reexamined," *Systematic Zoology* 40 (1991): 335-63; Milford Wolpoff and Alan Thorne, "The Case against Eve," *New Scientist* (22 June 1991): 37-41; Alan R. Templeton, "The 'Eve' Hypothesis: A Genetic Critique and Reanalysis," *American Anthropologist* 95 (1993): 51-72; Christopher Wills, "When Did Eve Live? An Evolutionary Detective Story," *Evolution* 49 (1995): 593-607; Alan R. Templeton, "Out of Africa? What Do Genes Tell Us?" *Current Opinion in Genetics and Development* 7 (1997): 841-47; Mark Stoneking, "Women on the Move," *Nature Genetics* 20 (1998): 219-20; Todd R. Disotell, "Human Evolution: Origins of Modern Humans Still Look Recent," *Current Biology* 9 (1999): 647-50.

Of the many books about O.J. Simpson, seemingly written by almost half the people involved in the case and trial, we consulted a few. Alan Dershowitz, one of Simpson's many lawyers, wrote *Reasonable Doubts: The O.J. Simpson Case and the Criminal Justice System* (New York: Simon and Schuster, 1996). *Without a Doubt* (New York: Viking, 1997), written by prosecutor Marcia Clark, with Teresa Carpenter, provides a view from the other side. Neither of these books pretends to be impartial. Michael

Lynch and Sheila Jasanoff edited a special issue of *Social Studies of Science* 28: 5-6 (1998) on forensic science in the wake of the Simpson trial, which contains a number of interesting perspectives.

Chapter 5:
Confronting Nature in Lab and Field

In recent years there have been many studies of laboratories and the work that is done in them. *Laboratory Life: The Construction of Scientific Facts*, 2nd ed., by Bruno Latour and Steve Woolgar (Princeton, NJ: Princeton University Press, 1986) is the result of Latour's quasi-anthropological study of the Salk Laboratory in California.

Life Among the Scientists: An Anthropological Study of an Australian Scientific Community, by Max Charlesworth, Lyndsay Farrell, Terry Stokes, and David Turnbull (Oxford: Oxford University Press, 1989), deals in greater depth with most of the topics covered in this chapter. Some other useful and interesting books on laboratories and experimentation are: Karin Knorr Cetina, *The Manufacture of Knowledge: An Essay on the Constructivist and Contextual Nature of Science* (Oxford: Pergamon Press, 1981); Sharon Traweek, *Beamtimes and Lifetimes: The World of High Energy Physicists* (Cambridge, Mass.: Harvard University Press, 1988); Carolyn Merchant, *The Death of Nature: Women, Ecology, and the Scientific Revolution* (San Francisco: Harper & Row, 1980); and Martin H. Krieger, *Doing Physics: How Physicists Take Hold of the World* (Bloomington: Indiana University Press, 1992).

Steven Shapin and Simon Schaffer's *Leviathan and the Air-Pump: Hobbes, Boyle, and the Experimental Life* (Princeton, NJ: Princeton University Press, 1985) is the now-classic study of the solidification of experimentation as a scientific activity. Peter Dear's *Discipline and Experience: The Mechanical Way in the Scientific Revolution* (Chicago: Chicago University Press, 1995) provides one explanation for how artificial phenomena came to be recognized as a legitimate object of scientific study. Nobel Prize winner Roald Hoffman makes some points that are quite congruent with ours here in *The Same Not The Same* (New York: Columbia University Press, 1995). Hoffman's book has the added nice feature that it focuses on chemistry, which generally receives short shrift in discussions of science.

Much of the story of Daniel Simberloff and E.O. Wilson's Florida Keys experiment is contained in Wilson's autobiography, *Naturalist* (New York: Warner Books, 1995). An excellent account of the theory of island biogeography is David Quammen's *The Song of the Dodo: Island Biogeography in an Age of Extinctions* (New York: Scribner, 1996).

Nicholas Wade, in *The Nobel Duel: Two Scientists' 21-year Race to Win the World's Most Coveted Research Prize* (Garden City, NY: Anchor Press/Doubleday, 1981), tells the exciting story of the competition between Roger Guillemin and Andrew Schally for the Nobel Prize.

On objectivity: In the nineteenth century photographs sometimes displaced sketches as illustrations of anatomical phenomena, not because the photographs were better than the sketches— they usually showed less detail—but because they weren't subject to artists' preconceptions and adjustments. For an interesting discussion of this, see Lorraine Daston and Peter Galison's, "The Image of Objectivity," *Representations* 40 (1992): 83-128. Nonetheless, photographs are typically adjusted to make them more useful; this process is described by Michael Lynch in his "The externalized retina," in Michael Lynch and Steve Woolgar, eds., *Representation in Scientific Practice* (Cambridge, Mass.: MIT Press, 1990). June Goodfield's *An Imagined World: A Story of Scientific Discovery* (New York: Harper & Row, 1981) is the source of our story about Anna Brito.

Chapter 6:
Doing Science in the Real World

Barry Werth tells the story of Joshua Boger and his company Vertex Pharmaceuticals in *The Billion-Dollar Molecule: One Company's Quest for the Perfect Drug* (New York: Touchstone, 1994). We could have chosen any one of a number of other such episodes to illustrate the point that in commercial research, business and scientific decisions are inseparable. Another book that is instructive in this regard is Tracy Kidder's *Soul of a New Machine* (New York: Avon Books, 1981).

Bruno Latour and Steve Woolgar's *Laboratory Life*, mentioned above, gives a circulation model of scientific credibility. Robert Merton is responsible for a slightly different version of the same idea, under the heading of the "Matthew Effect," in *The Sociology of Science* (Chicago: University of Chicago Press, 1973). Jonathan Cole and Burton Singer use that framework to explain differences in productivity between women and men in "A Theory of Limited Differences: Explaining the Productivity Puzzle in Science." That essay is published in *The Outer Circle: Women in the Scientific Community* (New York: W.W. Norton, 1991), a useful book on women in science, edited by Harriet Zuckerman, Jonathan R. Cole, and John T. Bruer. Mary Frank Fox's "Gender, Environmental Milieu, and Productivity in Science" is published in the same volume. Natalie Angier's *New York Times* article "Women Swell Ranks of Science, but Remain Invisible at the Top" (21 May 1991): C1, C12, is a good snapshot on these issues and includes the figures we cite here. Londa Schiebinger's *Has Feminism Changed Science?* (Cambridge, Mass.: Harvard University Press, 1999) gives a more comprehensive overview and also discusses how feminism has affected the facts and theories of science.

Derek de Solla Price's classic, *Little Science, Big Science, ... And Beyond* (New York: Columbia University Press, 1986; 1st ed., 1963) contains a number of fascinating comments on the growth of science and its changing patterns of research. The story of cloud chambers and bubble chambers is told in Peter Galison's monumental history of instruments in particle physics, *Image and Logic: A Material Culture of Microphysics* (Chicago: University of Chicago Press, 1997). Daniel Koshland's discussion of the overwork of scientists is in "The Addictive Personality," *Science* 250 (1990): 1193.

Stanley Ewen and Arpad Pusztai published their results as "Effects of Diets Containing Genetically Modified Potatoes Expressing Galanthus Nivalis Lectin on Rat Small Intestine," *The Lancet* 354 (1999): 1353. Richard Horton's analysis of the Pusztai affair is "Secret Society: Scientific Peer Review and Pusztai's Potatoes," *Times Literary Supplement* (17 December 1999): 8-9. There are a number of letters and editorials on the subject in various 1999 issues of *The Lancet*.

Chapter 7:
The End of Science?

The two books entitled *The End of Science* are: first, journalist John Horgan's argument that there is a good chance that everything important has already been discovered, in *The End of Science: Facing the Limits of Knowledge in the Twilight of the Scientific Age* (New York: Broadway Books, 1996); and the twenty-fifth Nobel Conference volume edited by Richard Q. Elvee, *The End of Science? Attack and Defense* (Lanham, Md.: University Press of America, 1992).

"The Physicists' Debates on Unification in Physics at the End of the 20th Century," an article by Jordi Cat in *Historical Studies in the Physical and Biological Sciences* 28 (1998): 253-99, contains a number of references to physicists' recent discussions of reductionism. More popular discussions are in Roger Penrose's *Shadows of the Mind* (Oxford: Oxford University Press, 1994); Jacob Bronowski's *A Sense of the Future* (Cambridge, Mass.: MIT Press, 1978); and Ilya Prigogene's millennial contribution *The End of Certainty* (New York: Free Press, 1996). Steven Weinberg's *Dreams of a Final Theory* (New York: Vintage Books, 1992) is his book-length defence of reductionism in science; his article "Reductionism Redux" in *The New York Review of Books*, (5 October 1995): 39-42, is an excellent overview.

Philip Anderson has written a number of articles arguing against reductionism from his stance as a solid-state physicist. A representative one is "More is Different: Broken Symmetry and the Nature of the Hierarchical Structure of Science," *Science* 177 (1972): 393-96. Michael Polanyi's comments are in *Personal Knowledge* (London: Routledge & Kegan Paul, 1958), p. 382.

Daniel J. Kevles's *The Physicists: The History of a Scientific Community in Modern America*, 2nd ed. (Cambridge, Mass.: Harvard University Press, 1995) contains a good analysis of the debacle around the superconducting supercollider. James Krumhansl's comments about the rhetoric of the SSC's proponents is in his editorial "Unity in the Science of Phyiscs," *Physics Today* (March 1991): 33-38.

Lewis Wolpert makes his case for a cautious version of reductionism, similar to the one we advocate here, in "Do We Understand Development?" *Science* 256 (1994): 571-72. Alex Rosen-

and Philosophy 12 (1997): 445-70, picks up that case and discusses it with care.

Roald Hoffman's beautiful *Chemistry Imagined: Reflections on Science* (Washington: Smithsonian Institute Press, 1993) is a compelling argument for the human nature of scientific representations. Sergio Sismondo's *Science Without Myth: On Constructions, Reality, and Social Knowledge* (Albany, NY: State University of New York Press, 1996) makes that case in a drier fashion, but contains a number of other useful references.

The interview with the corporate biochemist comes from Paul Rabinow, *Making PCR: A Story of Biotechnology* (Chicago: University of Chicago Press, 1996).

Picture Credits

Front Matter

1 Nobeyama, Mm-wave Array, Japan, courtesy Judith Irwin.
2 Runge's coloured spheres, *Die Farbenkugel,* Hamburg, 1810.
3 Elementary particle track, CERN Library.

Chapter 1

1 Jean-Louis David, *The Lavoisiers,* Musée du Louvre, Paris.
2 Jean Perrin: Palais de la Découverte, Paris.
3 Pablo Picasso: Ermine Erscher, *Picasso Bon-vivant.*
4 Giovanni Alfonso Borelli, *De Motu Animalium:* Louise Darling Biomedical Library, University of California.
5 Albert Einstein: Einstein Collection, Hebrew University, Jerusalem.
6 Niels Bohr and Werner Heisenberg: Niels Bohr Archives, Copenhagen.
7 Stanley Kubrick, *2001: A Space Odyssey.* Reprinted by permission.

Chapter 2

1 Bohr atom: Niels Bohr Archives, Copenhagen.
2 Albert Einsten: Einstein Collection, Hebrew University, Jerusalem.
3 Marie and Pierre Curie: Palais de la Découverte, Paris.
4 Stereoscope: Bibliothèque Nationale, Paris.
5 Paul Cézanne, *Still Life with Basket of Fruit:* Musée du Louvre, Paris.
6 Ernest Rutherford: Cavendish Laboratory, Cambridge, UK.
7 Eadward Muybridge, *Daisy Jumping a Hurdle:* Palais de la Découverte, Paris.
8 Etienne-Jules Marey, *Walking Horse:* Bibliothèque Nationale, Paris.
9 Marcel Duchamp, *Nude Descending a Staircase:* Philadelphia Museum of Art. © Estate of Marcel Duchamp / SODRAC (Montreal) 2001.
10 Frantisek Kupka, *Woman Picking Flowers:* Musée national d'art moderne, Paris. © Estate of F. Kupka/SODRAC (Montreal) 2001.
11 Giacomo Balla, *Young Girl Running on a Balcony:* Balla Collection, Rome. © Estate of G. Balla / SODRAC (Montreal) 2001.
12 Giacoma Balla, *Study for a Young Girl:* Balla Collection, Rome. © Estate of G. Balla / SODRAC (Montreal) 2001.
13 Giacomo Balla, *Dynamism of a Dog on a Leash:* Albright Knox Art Gallery, Buffalo / SODRAC (Montreal) 2001.
14 Piet Mondrian, *Composition C:* Museum of Modern Art, New York. © BEELDRECHT / SODART 2001.
15 Piet Mondrian, *Broadway Boogie-Woogie:* Museum of Modern Art, New York. © BEELDRECHT / SODART 2001.
16 Casimir Malevich, *The Fourth Dimension:* Museum of Amsterdam.
17 Werner Heisenberg: Niels Bohr Archives, Copenhagen.

Chapter 3

1 *Reason Reveals the Truth:* Denis Diderot, *L'Encyclopédie,* Bibliothèque National, Paris.
2 *Foucault's Pendulum in the Pantheon: L'Illustration* (Paris, 1851), The British Library.

3 Charles Darwin: Francis Darwin, *The Life and Letters of Charles Darwin* (London: John Murray, 1887).

4 Darwin's Finches: Palais de la Découverte, Paris.

5 Ernst Haeckel's radiolarians: John Romanes, *Darwin after Darwin* (London, 1892).

6 Frederic Brewster Loomis, *The Evolution of the Horse* (Boston: Marshall Jones Company, 1926).

7 William K. Gregory, *Our Face from Fish to Man: A Portrait Gallery of Our Ancient Ancestors and Kinsfolk together with a Concise History of Our Best Features* (New York: G.P.. Putnam's Sons, 1929).

8 Honoré Daumier, *L'amateur d'estampes*: Glasgow City Museum, Glasgow.

9 Denis Diderot, *L'Encyclopédie*, Bibliothèque Nationale, Paris.

Chapter 4

1 Sébastien Le Clerc, *L'étude de la zoologie au Jardin des Plantes*: Prints Collection, Bibliothèque Nationale, Paris.

2 DNA photograph: Lawrence Berkeley National Laboratory.

3 Out of Africa: *Science* 257 (14 August 1992), 873.

4 Gustave Doré, *Eve and Adam*: Robert Vaughan, ed., *Milton's Paradise Lost* (New York: Collier, n.d.).

5 *Philosophical Transactions of the Royal Society* 3 ("Munday, May 8, 1665"): 33-34.

6 M.A. Sutton, J.K. Schjorring, and G.P. Wyers, "Plant-atmosphere exchange of ammonia," *Philosophical Transactions fo the Royal Society of London* A (1995): 271.

7 The Difference Engine: Henry Provost Babbage, *Babbage's Calculating Engines* (London: Spon & Company, 1889).

8 Honoré Daumier, *Une Cause Criminelle*, J. Paul Getty Museum.

Chapter 5

1 Robert Boyle, *A Continuation of New Experiments Physico-Mechanical Touching the Spring and Weight of the Air*, 1669. Houghton Library, Harvard University.

2 *In Liebig's Laboratory*, Bibliothèque Nationale, Paris.

3 Microphotographic apparatus: S.T. Stein, *Das Licht* (Halle, 1886).

4 E.S. Horning and G.M. Findlay, "Microincineration Studies of the Liver in Rift Valley Fever," *Journal of the Royal Microscopic Society* 54 (1934): 9-16.

5 Darwin's beetles: Palais de la Découverte, Paris.

6 Mme. Lavoisier's sketch of respiration experiments: *Traité Elementaire de Chimie* (Paris: E. Grimaux, 1888).

7 The Nobel Prize: The Ava Helen and Linus Pauling Papers, Special collections, Oregon State University.

8 Mice strains: Sergio Sismondo, after *Natural History of New York, Part I*. (New York: D. Appleton & Co., 1842).

9 Lawrence's accelerator: Lawrence Berkeley National Laboratory.

10 Hilac Accelerator: Lawrence Berkeley National Laboratory.

Chapter 6

1 Von Edelfeldt, *Pasteur in his Laboratory in the Ecole Normale Supérieure* (1885): Musée Pasteur, Paris.

2 Sébastien Le Clerc, Louis XIV visiting the Academic des Sciences in 1671. Print Collection, Bibliothèque Nationale, Paris.

3 Nicola Testa. Wellcome Institute Library, London.

4 Benjamin Franklin, from *A History of Electricity* (1752): Palais de la Découverte, Paris.

5 Luis Alvarez and Bubble Chamber: Lawrence Berkeley National Laboratory.

6 Christo, *Wrapped Reichstag*: Taschen Library, Berlin.

7 *Mother and Daughter Using Microscope*, from Abbé Jean-Antoine Nollet, *Leçons de la Philosophie Experimentale* (1748), Bibliothèque Nationale, Paris.

8 Sébastien Le Clerc, *The Academy* (1698): Print Collection, Bibliothèque Nationale, Paris.

9 Fritz Lang, *Metropolis*. Reprinted by permission.

Chapter 7

1 Francis Bacon, *Great Instauration* (1620), Science Library, Ottawa.

2 Pablo Picasso, *Le Chef d'oeuvre inconnu*: Musée Picasso.

3 DNA model: Lawrence Berkeley National Laboratory.

4 Mount Wilson Telescope, Library of Congress.

5 White House Briefing. Reprinted with permission from *Physics Today* (March 1987): 47. © 1987, American Institute of Physics.

6 Sudbury Neutrino Observatory. Collaboration painting by Garth Tietjen.

7 Frantisek Kupka, *Newton's Discs*, Musée national d'art moderne, Paris. © Estate of F. Kupka / SODRAC (Montreal) 2001.

8 Piet Mondrian, *The Red Tree*: Gemeentemuseum, The Hague. © BEELDRECHT / SODART 2001.

9 Piet Mondrian, *Tree*, Gemeentemuseum, The Hague. © BEELDRECHT / SODART 2001.

10 Piet Mondrian, *The Gray Tree*: Gemeentemuseum, The Hague. © BEELDRECHT / SODART 2001.

11 Elementary particle track, CERN Library.

12 Hans Hartung, *Peinture*. © Estate of H. Hartung / SODRAC (Montreal) 2001.

13 Vassily Kandinsky, *Improvisation*. Russian Museum, St. Petersburg.

14 PET scan: Lawrence Berkeley National Laboratory.

15 CERN: Lawrence Berkeley National Laboratory.

16 Pieter Breugel, *Tower of Babel* (detail): Kunsthistorisches Museum, Vienna.

17 Sébastien Le Clerc, *Astronomie*, Bibliothèque Nationale, Paris.

18 Space Telescope: Science Institute Library.

[The Authors of this book and the Publisher have made every attempt to locate the authors of copyrighted material or their heirs or assigns, and would be very grateful for information that would allow them to correct any errors or omissions in a subsequent edition of the work.]

Acknowledgements

This book has benefitted from the assistance of many individuals and organizations. We are particularly indebted to Stephen Anderson, Elizabeth Davey, Cathleen Hoeniger, Bruce Laughton and Penny Roantree. Thanks also to Barbara Conolly and Michael Harrison at Broadview Press, Pat Valentine at True to Type for her elegant design and layout of the book, Katia Macias-Valadez at SODART, Stephanie Rochette at SODRAC and Steve Sheffield at Fourway Imaging. Finally, special thanks to Barbara Castel and Laura Murray for their insight and encouragement.

DATE DUE